Biotechnology, Agriculture, and Food Security in Southern Africa

Biotechnology, Agriculture, and Food Security in Southern Africa

Edited by Steven Were Omamo and Klaus von Grebmer

International Food Policy Research Institute
2033 K Street, N.W.
Washington, D.C.

Food, Agriculture, and Natural Resources Policy Analysis Network (FANRPAN)
12th Floor Social Security Centre
Corner J. Nyerere Street and Sam Nujoma Avenue
Harare, Zimbabwe

Copyright © 2005 International Food Policy Research Institute

All rights reserved. Sections of this material may be reproduced for personal and not-for-profit use without the express written permission of but with acknowledgment to IFPRI. To reproduce the material contained herein for profit or commercial use requires express written permission. To obtain permission, contact the Communications Division <ifpri-copyright@cgiar.org>.

International Food Policy Research Institute
2033 K Street, N.W.
Washington, D.C. 20006–1002
U.S.A.
Telephone +1–202–862–5600
www.ifpri.org

How to cite this book: Steven Were Omamo and Klaus von Grebmer, eds., Biotechnology, Agriculture, and Food Security in Southern Africa (Washington, DC, and Harare: IFPRI and FANRPAN, 2005).

Library of Congress Cataloging-in-Publication Data

Biotechnology, agriculture, and food security in Southern Africa / Steven Were Omamo and Klaus von Grebmer (editors).
    p.   cm.
Includes bibliographical references and index.
ISBN 0-89629-737-3 (alk. paper)
  1. Agricultural biotechnology—Government policy—Africa, Southern.
2. Transgenic organisms—Government policy—Africa, Southern.
3. Genetically modified foods—Government policy—Africa, Southern.
4. Food supply—Government policy—Africa, Southern. I. Omamo, Steven Were. II. Grebmer, Klaus von.
S494.5.B563B535   2004
664'.00968—dc22   2005005910

Contents

List of Tables vii

List of Figures ix

Foreword xi

Acknowledgments xiii

Introduction 1
Steven Were Omamo and Klaus von Grebmer

Chapter 1 Agricultural Biotechnology in Southern Africa: A Regional Synthesis 13
Doreen Mnyulwa and Julius Mugwagwa

Chapter 2 Consensus-Building Processes in Society and Genetically Modified Organisms: The Concept and Practice of Multistakeholder Processes 37
David Matz and Michele Ferenz

Chapter 3 Agricultural Biotechnology, Politics, Ethics, and Policy 71
Julian Kinderlerer and Mike Adcock

Chapter 4 Food Safety and Consumer Choice Policy 113
David Pelletier

Chapter 5 Biosafety Policy 157
Unesu Ushewokunze-Obatolu

Chapter 6 Intellectual Property Rights Policy 173
Norah Olembo

Chapter 7   Trade Policy   187
Moono Mupotola

Chapter 8   Lessons and Recommendations   199
Klaus von Grebmer and Steven Were Omamo

Appendix A   Workshop Proceedings for the FANRPAN-IFPRI Regional Policy Dialogue on Biotechnology, Agriculture, and Food Security in Southern Africa   223
Jenna Kryszczun and Steven Were Omamo

Appendix B   Workshop Program and Steering Committee Meeting Notes   271

Contributors   279

Index   281

Tables

1.1 Status of development and use of biotechnology techniques in Southern African Development Community countries, 2002   16

1.2 Status of development and use of biosafety systems in Southern African Development Community countries, April 2003   20

1.3 Levels of biotechnology awareness and public awareness strategies in Southern African Development Community countries, March 2003   27

1.4 Strengths, weaknesses, opportunities, and threats analysis of public awareness and public participation in southern Africa, November 2002   31

3.1 Agencies responsible for approval of commercial biotechnology products under the U.S. Coordinated Framework for the Regulation of Biotechnology   76

4.1 Contextual differences, United States and southern Africa   117

4.2 Key events in the development of agricultural biotechnology policy, 1973–2002   119

4.3 The effectiveness of FDA regulations in addressing various categories of concerns in transgenic plants   130

4.4 Unintended effects of genetic engineering breeding as of 2001   133

4.5 Outcomes and uncertainties of genetic modification under GM and non-GM policy options   148

5.1 Draft of proposed policy development framework for biosafety in the Southern African Development Community   162

6.1 Status of biosafety regulations and biotechnology policies or laws in eastern and southern Africa, 2004   174

6.2 Status of laws on intellectual property rights (IPR) in southern Africa, 2004   175

6.3 Participation of southern African countries in various intellectual property agreements, 2004   178

7.1 Production of and trade in genetically modified agricultural food products, 2000   188

7.2 Estimated percentage of international trade in genetically modified organisms, 2000   188

7.3 Fast-growing agricultural product areas under the African Growth and Opportunity Act   193

A.1 Emerging priority policy issues   259

A.2 Biotechnology development for food security   260

Figures

1.1 Gradient of biotechnologies in Southern African Development Community countries in terms of complexity and costs, 1993   15

2.1 How to conduct a conflict assessment   43

2.2 Key steps in the joint fact-finding process   45

2.3 The consensus-building process and the role of joint fact-finding   46

2.4 Phases of building agreement   49

3.1 European attitudes toward six applications of biotechnology, 2002   83

3.2 European optimism about technologies, 1991–2002   84

4.1 Cause and effect relationships involved in the introduction of *Bacillus thuringiensis* maize as a food for a human population   141

4.2 The relationship between scientific and normative (unscientific) dimensions of regulatory frameworks   145

Foreword

The role of modern biotechnology in spurring agriculture-led economic transformation and sustainable development in Africa is subject to furious scientific debate and intense public controversy. African governments therefore face enormous uncertainty and pressure as they deliberate on national and regional policies, programs, and regulations that attempt to maximize the benefits and minimize the risks of biotechnology products.

IFPRI does not imagine that it can bring resolution to these disagreements. Rather, as an international research organization with a mandate to identify policy solutions to hunger and poverty, IFPRI sees a need, and more importantly an opportunity, to help its partners. In particular, IFPRI sees the possibility that the heated debate on biotechnology in Africa might benefit from formal consensus-building platforms of the kind that have been effective in other parts of the world on controversial issues. Keen to ensure as neutral a process as possible, IFPRI committed its own resources to kick-starting the process of building such a consensus.

At about the same time that IFPRI was deliberating on its response to the challenging debate in Africa, the Harare-based Food, Agriculture, and Natural Resources Policy Analysis Network (FANRPAN) was also being approached by regional governments for help in increasing awareness about the range of policy issues raised by biotechnology in southern African agriculture. The Council of Ministers of Food, Agriculture, and Natural Resources of the Southern African Development Community had just established a subregional advisory committee on biotechnology and biosafety. FANRPAN had been involved in a process of reviewing biotechnology and biosafety policies and clearly saw the need for awareness building about biotechnology in the region.

Based on a memorandum of understanding signed in early 2003, and with technical support from the University of Massachusetts (Boston) Dispute Resolution Program and the Boston-based Consensus Building Institute, IFPRI and

FANRPAN embarked on a multistakeholder process of participatory awareness raising, joint fact-finding, and negotiation toward consensus on biotechnology, agriculture, and food security in southern Africa. The initiative's distinguishing feature was its explicitly process-based perspective within a framework involving many stakeholders. This feature distinguished it from other efforts in Africa with similar aims, most of which were episodic and lacked a clear conceptual framework.

A carefully managed but highly participatory process was planned, involving high-level policymakers, senior representatives of a range of stakeholder agencies, and respected scientific leaders, brought together for an integrated series of round-table discussions on biotechnology, agriculture, and food security in southern Africa. The first of three interlinked policy dialogues took place in Johannesburg, South Africa, on April 25–26, 2003. Following the Johannesburg meeting, the initiative evolved into a continent-wide effort known as the African Policy Dialogues on Biotechnology (APDB), a joint initiative between IFPRI and the Science and Technology Forum of the New Partnership for Africa's Development. With additional funding from the Rockefeller Foundation's Global Inclusion Program, a second dialogue took place in Harare, Zimbabwe, on September 20–21, 2004, under the auspices of the APDB initiative. A third dialogue is planned for 2005.

This volume comprises papers prepared as input to the first dialogue. In selecting topics for background papers, IFPRI and FANRPAN noted that the appearance of agricultural biotechnologies meant that governments were required to make new and unfamiliar choices in five areas: intellectual property rights, biosafety, trade, food safety and consumer choice, and public research. IFPRI and FANRPAN also noted the need for clarity on how political, ethical, and social imperatives interact within the context of agricultural biotechnology, and the implications for policy choice. Chapters analyzing policy issues in these seven areas, along with two synthesis chapters by the editors, result in a book that should be of interest to a wide range of individuals and organizations charged with making and shaping agricultural biotechnology policy in Africa.

Biotechnology offers important opportunities to African farmers and poor consumers. But biosafety policies need to be in place in order to move forward to responsible technology utilization. Most importantly, African policymakers need to be in a position to make their own well-informed decisions on the issues. IFPRI and FANRPAN are working toward these objectives. This collection of contributions represents an important step along the way to ensuring that biotechnology policies can facilitate increased food and nutrition security on the continent.

Joachim von Braun
Director General, IFPRI

Lindiwe Sibanda
Chief Executive Officer, FANRPAN

Acknowledgments

We would not have been able to complete this book without the dedication of each of the chapter authors. We are especially thankful for their patience and good humor, as we foisted one after another unreasonable editing and publication deadline on them. Beverly Abreu pulled together the first full version of the manuscript, also under great pressure, and also with good humor and skills. We thank her deeply. We are greatly indebted to Uday Mohan of the IFPRI Communications Division for his skillful and sensitive handling of the editing process. Our thanks go to Joel Cohen of IFPRI for his helpful review of the first full draft of the manuscript; his excellent comments and suggestions for improvement came at a crucial time. Two external reviewers also provided useful suggestions for improvement. We thank Joachim von Braun, IFPRI's director general, for having the vision and courage to commit to this initiative; Tobias Takavarasha, former chief executive officer (CEO) of FANRPAN; and Lindiwe Sibanda, the current CEO, for sharing that vision and commitment. Finally we thank all the participants in the Johannesburg policy dialogue for their input, support, and dedication.

Introduction

Steven Were Omamo and Klaus von Grebmer

Biotechnology disputes fall into the ever-expanding category of policy disputes characterized by multidimensionality and complexity. By their very nature, these disputes are centered around politically charged issues of allocation of rights to resources, as well as distribution of the benefits and costs of technological change. They typically involve a high degree of scientific uncertainty, long time horizons, and decisionmaking at multiple jurisdictional levels. Such disputes are therefore likely to pose exacting challenges. They involve a wide range of political, economic, social, and scientific considerations. Their satisfactory resolution therefore requires multistakeholder participation in a process of finding and maintaining a dynamic balance between political and technical priorities. In this process civil society can provide much of the expertise and creative thinking that is required to identify needs, generate innovative policy options, and implement agreements while governments retain their preeminent functions of ultimate decisionmaking.

At the beginning of 2003, the International Food Policy Research Institute (IFPRI) and the Food, Agriculture, and Natural Resources Policy Analysis Network (FANRPAN) embarked on a multistakeholder initiative aimed at raising awareness, promoting dialogue, and catalyzing consensus-building mechanisms toward improvement of the institutions and policies governing biotechnology in agriculture and its implications for food security in southern Africa.

The primary motivation for the initiative was the food emergency facing southern Africa. Inadequate, poorly timed, or inappropriate policy responses to small domestic food supplies combined with inadequate human, infrastructural, and organizational capacity in domestic markets to leave millions of people in the region at risk of starvation. Several years ago, in 1991, similar interactions among poor weather, policy failures, and market failures left millions of southern Africans similarly exposed. But the food emergency of 2002–03 was different from that of 1991–92 in one crucial respect. Thousands of tons of food available to help cover shortages in southern Africa contained unspecified amounts of genetically modified (GM) grain (specifically, *Bacillus thuringiensis* [*Bt*] maize) and were thus considered suspect—or even poisonous—by some governments unsure of the implications of GM food for human health and the environment. Efforts to accommodate that uncertainty pitted erstwhile partners in national and regional food relief against one another in an increasingly heated political environment.

The presence of GM food in the region not only raised political temperatures; it also rendered inordinately more difficult a range of basic tasks and operations in food relief—for example, moving grain through ports and across borders. Perceived risks associated with GM food created an entirely new set of transaction costs. How, for instance, was Malawi to move maize donated by the United States, and thus containing *Bt* maize, through Tanzania in mid-2002 in the absence of complementary biosafety protocols in Tanzania and Malawi, and in the absence of associated testing machinery? Ad hoc measures had to be hammered out, under extreme pressure, on such seemingly mundane issues as how to load grain into rail cars and trucks with minimal "escape," how to cover the loaded cars and trucks, and how long to allow the loaded cars and trucks to sit in given positions. The opportunity cost associated with such logistical hurdles, coupled with the region's general reticence toward potentially life-saving but GM food, elicited intense scrutiny and opprobrium from food donors and relief agencies.

Countries in the region have responded to the debate on genetically modified organisms (GMOs). At a meeting of the Southern African Development Community (SADC) Council of Ministers for Food, Agriculture, and Natural Resources (FANR) on July 5, 2002, in Maputo, Mozambique, it was noted that the lack of a harmonized (regional) position on GMOs was creating serious operational problems in movement of food and nonfood items. Consequently, the council advised member states to engage in bilateral consultations and to explore mechanisms to facilitate movement of humanitarian aid in the form of food that might contain GMOs. The FANR ministers approved the establishment of an advisory committee on biotechnology and biosafety to develop guidelines to safeguard member states against potential risks of GMOs in the areas of trade, food safety, contamination of genetic resources, ethics, and consumer concerns (SADC 2003). The committee has been constituted and is developing the requested guidelines.

More broadly, African leaders have resolved to build regional consensus and strategies to address concerns emerging with advances in modern biotechnology, including genetic engineering. This resolution is manifested in decisions of the

African Union (AU) and the New Partnership for Africa's Development (NEPAD). Specifically, Decision EX.CL/Dec. 26 (III) of the AU Summit calls for the development of a common African position on biotechnology. Those attending the second meeting of the NEPAD Science and Technology Steering Committee decided that the Secretariat of NEPAD and the AU Commission should establish a high-level panel of experts to prepare a comprehensive African strategy and a common position on biotechnology, including applications for agriculture, health, the environment, mining, and manufacturing. This high-level panel will be comprised of eminent experts and opinion leaders who will provide comprehensive advice on current policy issues associated with the ethical, social, regulatory, economic, scientific, environmental, and health aspects of biotechnology, including genetic engineering.

Clearly the content and nature of the debate on how to respond to food crises have been fundamentally and irreversibly altered. So too have been those elements of the debate on how to achieve longer-term agricultural growth and food security through self-sustaining processes of growth fueled by technological advance in agriculture. Many stakeholders believe that in the wake of GM food will come GM agricultural technologies. Enduring uncertainties and controversies over the relevance, efficacy, sustainability, and safety of those technologies appear to render such a progression unpalatable to many.

A key recognition is that the uncertainties and controversies surrounding the role of biotechnology in agricultural development and food security enhancement are not confined to southern Africa but are global in scope. In most cases these uncertainties and controversies appear to have two dimensions. One dimension applies to relatively well-informed stakeholders, the other to relatively uninformed stakeholders. Because the relatively uninformed, either by design or by default, often rely on the relatively well-informed for guidance, understanding the foundations of differences among informed stakeholders is crucial. The problem becomes even more complex when there are grave discrepancies among the relatively well-informed (in the United States and the European Union) on how to proceed and when these stakeholders try to persuade the relatively uninformed to follow their respective lines of reasoning in dealing with this technology. Multistakeholder dialogues help to convey information on all aspects of certain issues and thus contribute to informed and democratic choices.

Conflicting Disciplinary Perspectives: Biophysical Sciences vs. Social Sciences vs. Humanities

Differences among informed stakeholders in the debate on biotechnology in agriculture appear to stem in part from contrasting disciplinary approaches and methodologies in knowledge generation. The tight, narrow, experiment-based hypothesis-testing approaches in the biophysical sciences contrast with those in the social sciences, which are concerned with looser, broader collective behavioral hypotheses in which both theory and data provide ambiguous guidance on causal relationships. Increasing use of experimentation in the social sciences holds prospects for bridging this particular disciplinary divide. But it reinforces another, namely that between the sciences on the one hand and the humanities on the other. The reductionism that drives model building and hypothesis testing in the sciences is negated in the humanities, where explanation is often built on narrative depictions of dialectic tensions between individual agency and societal determinism.

Consider the following hypothetical exchange among a biophysical scientist, a social scientist, and a scholar from the humanities—say, a molecular geneticist, an economist, and a social historian. Suppose they are discussing the value of research on how resistance to trypanosomosis (a dominant parasitic livestock disease in Africa) might be maintained and enhanced while retaining and reinforcing characteristics of economic importance to farmers, and on how "trypanotolerance" can be imparted to susceptible animals while retaining their other important traits. Historically this research has been field-based, but it is increasingly biotechnology-driven.

> *Molecular geneticist:* This research is extremely valuable to Africa. The techniques we employ are state-of-the art. We can demonstrate that marker-assisted selection of target genes within breeds of disease-tolerant animals, and marker-assisted introgression of target genes from tolerant to susceptible breeds will give rise to productivity gains due to increased capacity to control parasite development and thus limit the onset of anemia. The impacts on livestock health and thus on poverty alleviation in Africa will be enormous.

> *Economist:* Yes, but how sure are you about those productivity gains? When will they appear, and with how much variability? Remember that farmers are pretty conservative in their breed preferences, particularly those farmers rearing multipurpose animals in mixed crop-livestock production systems, as in much of Africa. Outputs of your research must meet farmers' needs. Those needs are reflected in selections of animals based on traits for which heritability is already known. Not all of those traits are linked to trypanotolerance. The relevance of research on trypanotolerance, and, most important, the likelihood that farmers will actually adopt outputs of that research and realize the potential gains are therefore not at all clear.

> *Social historian:* The history of the last two centuries is replete with examples of new and revolutionary technologies. That history teaches that

although many of these inventions did change the world for the better, many did not. Most important, a significant number of these technologies turned out to have both benefits and risks that were wholly unanticipated beforehand. In many cases, some benefits and risks were not discerned until long after the technologies were well entrenched. And all along there were heated arguments for and against this or that technology. If there is a lesson from this, it is that only time and a commitment to openness in identifying and debating both benefits and costs will bring increased understanding of what this kind of technology might mean to Africa's livestock keepers.

Competing Paradigms: Modernism and Postmodernism

The deep epistemological divergences defined by alternative disciplinary perspectives are further accentuated by a more fundamental conceptual (paradigmatic) clash based on differences surrounding the role of science and technology in human development. That clash pits modernists against postmodernists.

Modernism is predicated on beliefs that science and technology yield outcomes that are largely positive and beneficial, and that with scientific and technological advance human progress and development are inevitable and good. For modernists human history is captured in global, culture-neutral theories and patterns ("metanarratives") in which levels and rates of scientific and technological advance are decisive, and in which agency (and thus power) resides primarily with countries and peoples occupying prominent positions on scientific and technological frontiers.

Postmodernism is largely a reaction to the assumed certainty of scientific, or objective, efforts to explain reality. For postmodernists reality is constructed, knowledge is subjective, and thus interpretation is everything. Progress and development are far from being natural outcomes of scientific and technological advance, or of human history. Rather, the only sure outcome of science and technology, and of the passage of time, is change. Concrete experience therefore takes precedence over abstract principles, implying multiple ways of knowing, multiple truths, multiple sources of agency (and power), and a general incredulity toward metanarratives. According to this schema, science and technology have had their chance, but have failed to deliver. Scientists can no longer stand apart from society, unwilling to share the burden of finding solutions to the risks imposed by their inventions.

Consider the following hypothetical exchange between a modernist and a postmodernist on the risks posed by GM technologies.

Modernist: There is far too much woolly, antiscientific thinking flying around. Prove to me that GM technologies pose any more risk than do traveling in a car or flying in a plane. The risks posed by GM crops are dwarfed by the risks we confront every day, using conventional technologies. Just think about the risk of *not* taking advantage of the benefits promised by GM technology. Isn't that risk pretty clear? Isn't it continued hunger and poverty around the world? Isn't that outcome fully avoidable? Why *not* give Nature a nudge toward greater efficiency? Who are we to deny millions of poor, starving people the opportunity to live better, longer, more rewarding lives? What kinds of leaders would allow their citizens to suffer in that way?

Postmodernist: Not even the greatest scientist on this earth could "prove" that to you. You are enamored with science, yet you misapply it. You are blinded by it. The fact is that genetic engineering can unleash forces more powerful than even atomic energy, with unparalleled potential to harm life as we know it—and for all future generations. We also have a responsibility to these future generations. And those leaders you condemn out of hand—how can you begin to pass judgment on them when you have no idea about the political pressures they are facing? Who are you to impose your priorities and values on them?

Divergent Political Myths: South vs. North

A third divisive force in the debate on biotechnology in agriculture relates to political mythmaking—that is, to differences in myths about the nature of the global political order dominant in the South versus those dominant in the North.

In the South, a significant thread of political mythmaking springs from centuries of technology-driven domination by the North. It is therefore not uncommon to hear sentiments such as the following: "We must be constantly on guard against new forms of exploitation. This biotech thing is just another way for these people to make themselves richer—to make us more dependent on them. And if the Europeans and Americans want to fight over who will get richer from biotechnology, then they should not use us as proxy battle grounds."

In the North, despite sustained efforts toward greater inclusion and participation of "Southern" voices in development policy formulation, elements of the famous "white man's dilemma" persist. And so one might hear statements such as this: "We cannot turn our backs on millions of hungry people. Our future is intimately tied up with theirs. Luckily we have answers to their problems. The challenge we face lies in helping them—in helping their leaders—make the right choices."

Key elements of these clashes in disciplinary, paradigmatic, and political perspectives can be found in almost every public utterance on the role of biotechnology in agriculture. Not surprisingly, such elements run through and underpin the deepening controversy surrounding the role of GM food in meeting southern Africa's food shortage. They also hold sway in the debate on the role of biotechnology in meeting the region's longer-term agricultural growth and food security goals.

Objectives

There is an urgent need for greater clarity in concepts, facts, and potential actions toward the development of consistent institutions and policies governing biotechnology in southern African agriculture. Specifically, there is a pressing need to increase awareness, promote dialogue, and catalyze consensus-building mechanisms among national and regional stakeholder groups spanning public bodies (including parliamentary and judicial organs), the private sector, and civil society. The objective of the proposed initiative is therefore to facilitate and guide such dialogue and mechanisms.

Anticipated Outputs

The proposed process of policy dialogue and consultation is expected to result in the following set of outputs:

1. increased understanding among key national and regional policymakers and shapers of major developments and applications in agricultural biotechnology in the region, including central gaps and priority constraints;

2. greater awareness of, dialogue about, and consensus among key national and regional policymakers and -shapers on central policy trade-offs associated with GMOs in southern African agriculture;

3. greater awareness of, dialogue about, and consensus among key national and regional policymakers and -shapers on alternative institutional and organizational arrangements governing biotechnology in agriculture, and the potential consequences for national and regional responses to food crises and chronic food insecurity;

4. consensus recommendations (ideally in the form of a resolution or declaration) to enhance the ability of national and regional policies, programs, and regulations governing agricultural biotechnology products to spur agricultural growth and food security while ensuring protection of human health and the environment; and

5. an action plan for investment toward strengthened institutions and policies governing biotechnology in southern African agriculture, including an agenda for regional research, capacity strengthening, and outreach.

Methodology

A number of initiatives with similar objectives and outputs have been undertaken in several parts of the world. Their conceptual foundation would appear to be a method known as technology assessment (TA) developed by the U.S. Congress in the 1970s. TA was a political investment aimed at giving members of Congress access to independent, objective, and competent information on scientific and technical issues. As a result, congressmen were able to appreciate a fuller set of implications of legislative projects. Political choices among viable alternatives were thus better informed. Since then, the concept of TA has evolved further, largely in developed countries outside the United States. Wider stakeholder participation has been incorporated to better integrate varying interests and values. This greater emphasis on participation has reinforced the political dimension of TA and offered potential for democratizing technology through the entry of previously excluded knowledge, needs, experiences, and values. Questions of power, influence, and responsibility now arise explicitly and are confronted (Daele et al. 1997; Australian Museum 1999; Calgary 1999; Nentwich 1999; Goven, 2001). Efforts with some of these features have been attempted in developing countries—for instance, in Africa (Thamy 2002) and in South America (REDBIO 2001).

A New Initiative

The initiative promoted by FANRPAN and IFPRI adapts and applies key elements of the TA approach. Specifically, a carefully managed but highly participatory process is envisaged involving 40 to 50 high-level policymakers, senior representatives of a range of stakeholder agencies, and respected scientific leaders, brought together for an integrated series of roundtable discussions on biotechnology, agriculture, and food security in southern Africa. Three interlinked roundtable gatherings are planned, spread out over several months. A steering committee (SC) was appointed at the first meeting, with membership drawn from among the invitees. The SC will determine format, content, and participation at the meetings, supported by a working group drawn from the convening institutions.

To ensure a nonbiased approach, FANRPAN and IFPRI carefully considered issues of funding and legitimacy when planning the workshop, and took the position that the workshop would be funded only by IFPRI resources, although there were indications that other donors would be willing to fund. A self-selected internationally composed board of trustees governs IFPRI, and the board's composition and governance structures are transparent and public. FANRPAN has a similarly legitimate governance structure. Dr. John Mugabe was asked to chair the session not only because he is a skilled moderator, but also because his participation and the participation of NEPAD gave the workshop an Africa-wide legitimacy. Once a structure had been established, the group could approach other donors and there would not be a problem with legitimacy.

Roundtable Meetings: Toward Consensus Recommendations

It was decided that the 40 to 50 participants in the roundtable discussions would comprise 30 to 40 stakeholders (including members of the SC), 5 to 10 speakers and technical or subject matter experts, and 5 to 10 organizers. Given these numbers, the aim of the meetings would not be to reach definitive conclusions but rather to foster broad participation and open debate on clearly defined questions under procedurally fair conditions.

The first meeting was crucial. The meeting, which took place on April 24–26, 2003, in Johannesburg, South Africa, drew high-level policymakers, senior representatives of a range of stakeholder agencies, and respected scientific leaders. The meeting was carefully managed and highly participatory, using concepts and practices of multistakeholder processes. Key challenges revolved around ensuring that all relevant parties were involved, accurate scientific information was made available, links with official decisionmaking bodies were promoted, and fairness and efficiency were recognized and embraced as evaluation criteria. Seven background papers were prepared as input into the meeting. Two of these papers—a regional synthesis paper and a paper on concepts and practices of multistakeholder processes—were presented and discussed. The other five papers—which addressed a range of policy issues raised by biotechnology—were not formally presented, but all the authors were present at the meeting and contributed to the discussions. Material from both categories of papers is included in this volume.

A second round of studies will be commissioned based on the outcome of the first meeting. Experts selected by the SC will complete these studies. Results of the second round of studies will be discussed at the second meeting, which again will be two to three days in duration.

A third round of studies will be commissioned based on discussions at the second meeting. Again, experts selected by the SC will complete these studies. The third and final meeting will be devoted to discussing results of the third round of studies, identifying consensus recommendations (that is, a resolution or declaration), and, if relevant, outlining an appropriate follow-on action plan.

Organization and Overview of the Book

The implementation of agricultural biotechnology for food and feed production stimulates considerable controversy the world over, with strongly conflicting views not only about the technology itself but also about the ethical questions involved. Both aspects are open to interpretation and frequently polarize opinions both within and across countries. Nevertheless, with food security a major world challenge—perhaps the greatest challenge for southern Africa—agricultural biotechnology offers significant potential to alleviate food insufficiency by providing crops targeted to particular environments.

Chapter 1 provides a synthesis of the current status of agricultural biotechnology in southern Africa. The SADC countries vary in the degree to which they have developed and applied biotechnology and the associated systems governing its use; this situation should be exploited to ensure that all countries attain a minimum level of technical and regulatory capacity, especially for monitoring the development and use of genetic modification technologies and their resulting products. It is crucial that countries recognize their interdependence in the context of the current global economy and the need to monitor the movement of materials across borders. Adequately equipping the general public, especially farmers, will go a long way in building self-monitoring mechanisms, which will complement efforts by regulatory authorities to limit the unintended spread of GM products. An informed society will also influence the national research agenda, thereby ensuring that the constrained research and development resources of countries in the region are used to address priority issues.

Chapter 2 presents the key conceptual issues inherent in processes involving multiple stakeholders. Fundamentally, multistakeholder processes aim to address the multidimensional, complex, and intrinsically politically charged issues associated with technological change, such as the allocation of rights to resources and the distribution of costs and benefits. Three examples of such processes are presented to illustrate the central arguments, the social and political context within which policy change is debated and implemented, and the mechanisms available to facilitate discourse and ultimately decisionmaking. Success in reconciling deeply held perspectives and arriving at consensus on future directions depends on the extent to which the following challenges are met: (1) involving relevant parties in discussions and negotiations, (2) expounding accurate scientific information, (3) making significant linkages to official decisionmaking, and (4) adopting fairness and efficiency as evaluation criteria.

Chapter 3 addresses the range of political and ethical issues raised by biotechnology. It may be argued that governments and the scientific community have a duty to ensure the responsible diffusion of technology. Some argue that the current situation requires that technology be introduced immediately to alleviate suffering, while others take a more cautious approach, arguing that the technology should be introduced only after risk-benefit assessments have been carried out and appropriate legislation and regulatory frameworks are in place. The chapter seeks not to determine an answer but rather to put forward the issues and arguments to facilitate informed decisionmaking for each country.

Chapter 4 focuses on food safety and consumer choice policy, aiming to identify policy options and trade-offs relevant to southern Africa. In general, the genetic, metabolic, and food composition changes of future crops, including crops targeted to the needs of developing countries, are expected to make them more complex than first-generation crops and consequently may pose more complex regulatory questions. The chapter highlights the even greater scientific uncertainties in the southern African region, and proposes a scientific and values-based framework for analyzing policy options and trade-offs. A detailed analysis of U.S. Food and Drug Administration policies is also provided, including the scientific, legal, and political basis underlying them, to familiarize the SADC countries with the official position of the U.S. government as it relates to the United States and (to a large extent) international and bilateral discussions and negotiations.

Chapter 5 examines the role and purpose of biosafety, and the opportunities and challenges that the region faces regarding research and development in genetic engineering (GE) and the importation of GE products and their movement within and across SADC countries. Various positions are presented for exploration, again raising important issues of transboundary movement. The success of a biosafety policy framework will depend on country and regional commitment and cooperation, enabling policy instruments, sustainable human and financial support, and enhanced public understanding and awareness of biosafety issues and regional responses to the Cartegena Protocol.

Chapter 6 focuses on policy issues concerning intellectual property rights (IPR) in agricultural biotechnology, looking at both positive and negative aspects and considering urgent needs, including comprehensive policy guidelines for biotechnology application in southern African countries, IPR policies that define the role of protection in agricultural inventions, capacity development, partnerships among stakeholders to enhance technology transfer to address food security in southern Africa, networking and use of local groups in advocacy and awareness creation, and provision of the funding necessary to achieve these aims.

Chapter 7 addresses trade policy issues. As major food importers, the SADC countries must identify ways to take advantage of cheap GM grain while guarding against negative human health effects. Although there are advantages to the use of biotechnology, it is not a panacea for alleviating the area's food security needs. SADC member countries must act as a cohesive group in areas of mutual interest during negotiations of international agreements.

In the final chapter major lessons and recommendations are drawn, focusing on issues raised in expanding and sustaining multistakeholder processes in Africa, increasing awareness, and designing and implementing policy. Given the self-contained nature of the preceding chapters, readers interested principally in this set of issues can jump directly to this final chapter.

The proceedings of the April 2003 meeting in Johannesburg are found in Appendix A. The aim is not to provide a blow-by-blow account of the discussions, but rather to highlight the major issues addressed, the central areas of controversy and dispute, the key decisions made, and the most critical outcomes agreed to for future action. A central outcome of the meeting was the selection of the steering committee. The committee was selected so as to reflect the multistakeholder outlook of the dialogue. It was charged with preparing for future dialogues, facilitating linkages with other ongoing activities, and synthesizing and disseminating results of dialogues. Clusters of priority issues identified as a provisional list for the committee to consider for future dialogues fell into the following categories: biosafety policies and frameworks, trade, protection of intellectual property, risk assessment, protection and conservation of biodiversity, public and private sector roles, and policy formulation processes. The program and participant list are found in Appendix B.

References

Australian Museum. 1999. Lay panel report. First Australian Conference on Gene Technology in Food Chains, March 10–12, Canberra.

Calgary. 1999. "Citizen's Report on Food Biotechnology." Calgary Citizens' Conference on Food Biotechnology, March 5–7, Calgary.

Daele, W., A. Pühler, and H. Sukopp. 1997. *Transgenic herbicide-resistant crops: A participatory technology assessment. Summary report.* Discussion Paper FS 11 97-302. Wissenschaftszentrum Berlin für Sozialforschung.

Goven, J. 2001. Citizens and deficits: Problematic paths toward participatory technology assessment. Unpublished manuscript, University of Canterbury.

Nentwich, D. 1999. The role of participatory technology assessment in policy making. Paper presented at the Second EUROpTA Project Workshop, October 4–5, The Hague.

REDBIO. 2001. Declaration of Goiania. Declaration adopted by participants at the Fourth Latin American Meeting on Plant Biotechnology, June 4–8, Goiania.

SADC (Southern African Development Community). 2003. SADC responds to GMO debate. *SADC Seed Update* (electronic newsletter of the SADC Seed Security Network), issue no. 2 (January), http://www.sadc-fanr.org.zw/ssn/news/SADCSEEDUpdateN22003.pdf.

Thamy, R. 2002. Summary of presentations. Rockefeller Foundation / World Vision workshop on GMOs in African agriculture, May 14–16, Nairobi.

Chapter 1

Agricultural Biotechnology in Southern Africa: A Regional Synthesis

Doreen Mnyulwa and Julius Mugwagwa

The Convention on Biological Diversity defines biotechnology as "any technological application that uses biological systems, living organisms, or derivatives thereof, to make or modify products or processes for specific use." Defined this way, it clearly emerges that biotechnology is an old science, with many established uses in areas such as agriculture, medicine, forestry, mining, industry, and environmental management. The old applications are generally referred to as traditional biotechnology, and in agriculture these have been in use since the advent of the first agricultural practices for improvement of plants, animals, and microorganisms (Persley and Siedow 1999).

The application of biotechnology to agriculturally important crop species, for example, has traditionally involved the use of selective breeding to bring about an exchange of genetic material between two parent plants to produce offspring with desired traits such as increased yields, disease resistance, and enhanced product quality. The exchange of genetic material through conventional breeding requires that the two plants being crossed be of the same or closely related species.

The Generations of Biotechnology

The progress and development of biotechnology is generally divided into three broad categories, also referred to as generations of biotechnology. This acknowledges that biotechnology is not a new technology, but rather is a continuum of techniques and approaches that have evolved over time.

The first generation. This refers to the phase of biotechnology that was based on empirical practice, with minimum scientific or technological inputs. This phase stretched all the way from 12,000 BC to the early 1900s.

The second generation. Developments in fermentation technology, especially during the period between the two world wars, constitute what is generally referred to as the second generation or phase of biotechnology. Major products from this generation were antibiotics such as penicillin and other products such as vitamins and enzymes. Another critical event of this generation, beginning in the 1930s, was the development and use of hybrid crop varieties in the U.S. Corn Belt, which resulted in dramatic yield increases.

The third generation (new biotechnology). The third generation or phase of biotechnology, also referred to as the new or modern biotechnology, is the present one. A turning point occurred in 1953 with the discovery at Cambridge University (U.K.) of the structure of deoxyribonucleic acid (DNA), which is the molecular carrier of stored information. DNA is a long and winding molecule that is made up of a combination of several chemicals. Four related chemicals in DNA, called "bases," are lined up in specific sequences, and these specific sequences represent the information that determines the traits, features, characteristics, abilities, and functioning of cells within an organism.

The particular segment of DNA that contains information for a particular characteristic or trait is called a gene. In other words, the genes represent information that is passed on from one generation to the next. It is also important to point out that not all segments of DNA represent information that can be or is passed on from one generation to the next. Because DNA is made up of chemicals that are present in cells where many life-maintaining processes are occurring, the DNA needs to "protect" itself, and hence some segments of the DNA serve the purpose of ensuring that the DNA remains intact.

The Current Status of Biotechnology Research and Use in the SADC Region

Countries in the Southern African Development Community (SADC) region are employing various forms of biotechnological techniques in their agricultural, environmental management, forestry, medicine, and industry efforts, and have been since time immemorial. However, without doubt Africa is the region where biotechnologies are the least developed. There are many different explanations for this situation, but several schools of thought associate it with the perennial economic problems affecting the continent (Sasson 1993).

Figure 1.1 shows the gradient of biotechnologies in terms of complexity and costs. An analysis of the status of biotechnology in the different SADC countries will be presented and discussed based on this gradient.

Figure 1.1 Gradient of biotechnologies in Southern African Development Community countries in terms of complexity and costs, 1993

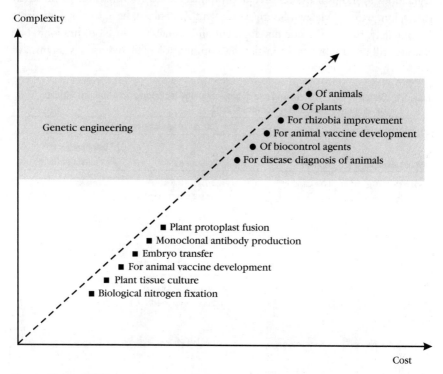

Source: Sasson 1993.

From studies conducted by the Biotechnology Trust of Zimbabwe (BTZ) in 2001 and 2002, and studies by other organizations such as the Rockefeller Foundation and International Service for National Agricultural Research, it can be seen that the main area in which biotechnology techniques are being applied in southern African countries is agriculture, with the major thrust being crop improvement. Techniques such as tissue culture are being applied in almost all the countries, mainly because of the less intensive nature of this technique in terms of human and infrastructural resources.

Modern biotechnological techniques, which include genetic engineering, are being employed in few of the countries, namely Malawi, South Africa, and Zimbabwe, and to a small extent in Mauritius and Zambia. Of all these countries, only South Africa has reached the commercialization stage insofar as products of genetic engineering are concerned. The rest are still at the laboratory research stage.

Tied closely to the issue of research is the development and implementation of regulations to monitor the research and products thereof. Only three countries in the region, namely Malawi, South Africa, and Zimbabwe, have legal mechanisms for biosafety, that is, the safe development and application of biotechnology. The rest are still at varying stages in the development of their biosafety systems. All

Table 1.1   Status of development and use of biotechnology techniques in Southern African Development Community countries, 2002

Techniques/category	Areas of application		
	Angola	Botswana	Democratic Republic of Congo
Tissue culture (TC)	Little is known	Used on a limited basis for root and tuber crops	Little is known
Genetic modification (GM)	Little is known	Limited research is being done at the University of Botswana. No field trials have been approved.	Little is known
Fermentation technology	Little is known	Used in the brewing industry	Little is known
Marker-assisted selection	Little is known	None	Little is known
Artificial insemination and embryo transfer	Little is known	Used in livestock breeding	Little is known
Molecular diagnostics and molecular markers	Little is known	Used on a limited basis in plant and animal disease diagnosis	Little is known
Biological nitrogen fixation	Little is known	Used mainly through integration of legumes in cropping systems	Little is known
Manpower training	Little is known	Training is offered in other natural science modules at the University of Botswana	Little is known

countries of the SADC region are signatories to the Cartagena Biosafety Protocol, an addendum to the Convention on Biological Diversity, which governs safe transboundary movement of living modified organisms, among other provisions for ensuring safety in biotechnology.

Table 1.1 gives details on the status of development and use of various biotechnological techniques in the southern African countries.

	Areas of application		
Lesotho	**Malawi**	**Mauritius**	**Mozambique**
Used in Irish potato production and micropropagation	Used in disease elimination and micropropagation for cassava, sweet potatoes, Irish potatoes, and horticultural crops	Used on a limited basis in sugar cane research	Used in cassava and Irish potato production, micropropagation, and disease elimination
None	At the research level for cassava improvement (virus resistance). *Bt* cotton trials have been conducted.	GM sugar cane is nearing field trials. Awaiting adoption of a biosafety framework.	None
None	Used for food and feed production	Widely used in the brewing industry	None
None	None	None	None
None	Used for cattle breeding	Used on a limited basis	None
None; serological techniques are still being used	At the research level for use in animal disease diagnosis and diversity studies	Serological techniques are still used for diagnosis	Serological techniques are still being used
Used for legumes only	Used for legumes only	Used for legumes	Used on a limited basis, for legumes
Undergraduate and graduate training is done in natural and agricultural science (National University of Lesotho)	Training is done in the natural and agricultural sciences (Bunda College of Agriculture). Most of the training is theoretical. No explicit biotech courses are offered.	No explicit biotechnology training is offered.	Limited training is done in the natural sciences and agriculture (Eduardo Mondlane University)

(*continued*)

Table 1.1 (continued)

Techniques/category	Areas of application		
	Namibia	Seychelles	South Africa
Tissue culture (TC)	Used in cassava and Irish potato production, micropropagation, and disease elimination	Little is known	Active programs have been developed employing TC techniques for root and tuber crops, ornamental and horticultural crops, and animal vaccine production
Genetic modification (GM)	None	Little is known	Most major universities and research institutions (both government and private) have major projects employing GM techniques. Both crops and animals are covered in the research activities. Insect-resistant cotton and maize and herbicide-tolerant cotton and soybeans are already being grown commercially.
Fermentation technology	Used in food processing (small-grain crops)	Little is known	Used widely in food and beverages as well as in pharmaceutical industries
Marker-assisted selection	None	Little is known	Used in maize and small-grains breeding as well as livestock research and development
Artificial insemination and embryo transfer	Used in cattle breeding	Little is known	Used in livestock research, breeding, and conservation
Molecular diagnostics and molecular markers	Serological techniques are still being used	Little is known	Used for plant and animal disease diagnosis
Biological nitrogen fixation	Used for legumes only	Little is known	Used for soil fertility improvement through legumes and inoculants
Manpower training	Limited training is done, but University of Namibia is currently pursuing setting up an MSc program in biotechnology	Little is known	Specific degree-level training programs are available at most major universities, with access to state-of-the art resources

Source: Mnyulwa and Mugwagwa 2002.

Areas of application			
Swaziland	**Tanzania**	**Zambia**	**Zimbabwe**
Used in Irish potato production and micropropagation	Techniques are employed relatively extensively for root and tuber as well as horticultural crops	Used in micropropagation and disease elimination for cassava, sweet potatoes, Irish potatoes, mushrooms, and planting materials	Used in micropropagation and disease elimination for sweet potatoes, mushrooms, Irish potatoes, and horticultural crops
None	Limited research is being done, e.g., on virus resistance in bananas. There have been no commercial releases, but trials on GM tobacco were conducted in 2002.	Use limited; still at the research level for cassava improvement (virus resistance). Confined trials of *Bt* cotton were conducted in 1999/2000.	Still at the research level, mainly for use in crop improvement for cowpeas, tobacco, maize, and sorghum. Confined trials of *Bt* maize and cotton have been conducted.
None	Used in the brewing industry and vaccine production	Used for food and feed production	Used in food processing, feed and vaccine production
None	Used in genetic characterization of coconuts, cashews, sweet potatoes, cassava, and coffee	None	At the research level for improvement of maize for drought resistance and for small-stock improvement
Used in cattle breeding	Used in livestock breeding and conservation	Used for cattle breeding	Used for cattle and small-stock breeding
Serological techniques are still being used	Used in plant and animal disease diagnosis	Used for plant and animal disease diagnosis and diversity studies	Used for plant and animal disease diagnosis and diversity studies
Used for legumes only	Used mainly for legumes; used on a limited basis for inoculants	Used for both legumes and inoculants	Used for soil fertility improvement for both legumes and inoculants
Training is done at the undergraduate level in natural sciences (University of Swaziland)	Training is done in agricultural and other life science courses. A BSc degree in biotech was recently introduced at Sokoine University. The country is also benefiting from the BIO-EARN (East African Regional Network on Biotechnology and Biosafety) program.	Training is done in the natural, veterinary, and agricultural sciences (University of Zambia). No explicit courses are offered in biotech.	Specific biotech training programs have been developed at both undergraduate and graduate levels (University of Zimbabwe, National University of Science and Technology, Africa University)

Biosafety Systems

An analysis of the SADC countries looking at the status of their development and use of policy systems to ensure the safe development and application of modern biotechnology shows that the countries are at different levels. They can be placed into three broad categories: those that have regulations, those that have draft regulations, and those that have yet to initiate or are still in the very initial stages of development of such regulations. Table 1.2 summarizes the countries' status.

Global and Regional Trends in the Production of GMOs

Worldwide it is estimated that more than 3 billion people have been consuming GM foods since their commercialization in 1996. The use of GM plant varieties

Table 1.2 Status of development and use of biosafety systems in Southern African Development Community countries, April 2003

Biosafety issue	Angola	Botswana	Lesotho
Status of development and implementation	There is no biosafety legislation at the moment. The Ministry of Agriculture has initiated discussions on biotechnology and biosafety issues.	There is no biosafety legislation in this country. A process to develop a national biosafety framework was initiated in 2002 with funding from the United Nations Environment Program (UNEP) and the Global Environment Facility (GEF). The National Coordinating Strategy Agency is the national focal point for biosafety.	A biosafety committee was set up in 2001 within the Environmental Protection Unit to initiate drafting of legislation. Very limited capacity for risk assessment
Use of biosafety system in regulation of work on or use of genetic engineering (GE)	It has been reported that GE grain imported by Namibia in 2001 was milled in Angola. Namibia s draft legislation guarded against contamination of the environment. Angola had and still has no regulations.	As indicated, there are no mechanisms in place to regulate GE and its products. The dependence of the country on agricultural produce from South Africa is a cause for concern.	There have not been any official reports of requests to conduct trials or import GM products. Absence of a biosafety system complicates the situation. However, some food products, especially from South Africa, are suspected to be GM.
Urgent requirements	Regulations, capacity building, public awareness	Development of a legal framework, capacity building, public awareness and participation.	Garnering support from policymakers, development of regulatory framework, capacity building, public awareness

represents the fastest adoption of a new technology according to reports of the International Service for the Acquisition of Agri-Biotech. The total land area devoted to cultivation of GM crops increased from 1.7 million hectares in 1996 to 52.5 million hectares in 2001 (James 2001). By 1998 some 40 new GM varieties were being cultivated worldwide, mainly in Argentina, Australia, Canada, China, France, Mexico, South Africa, Spain, and the United States.

The area of GM crops in the developing countries has increased over the years from 15 percent in 1998 to 25 percent in 2001, of which 22 percent was planted in Argentina and 3 percent in China. China is the only country where public researchers funded by the government produced and commercialized GMOs.

Malawi	Mauritius	Mozambique	Namibia
Has legally binding legislation on biosafety. A national biosafety committee was appointed, though the country has limited capacity for risk assessment.	Has a GMO bill that requires setting up a national biosafety committee (NBC)	Set up a committee within the Ministry of Environment to come up with interim legislation on biosafety. Legislation still being developed.	Has a national biosafety committee (the Namibian Biotechnology Alliance) and draft legislation. Also has very limited capacity for risk assessment.
An interim committee was consulted in the debate on whether Malawi should import GM food aid or not. Malawi accepted GM maize, with no conditions set.	Officially, no GE products have entered the country. The NBC is tasked with monitoring the registration and movement of GE products in the country. A locally developed GM sugar cane variety is awaiting release.	Has already officially received GM maize under the condition that it has to be milled before distribution to consumers. A framework is still needed to ensure effective monitoring of GM products.	Accepted milled GM maize in 2000. Rejected GM maize in 2002, and instead received food aid in the form of wheat, as per a recommendation by the national biosafety committee.
Raising awareness of new legislation among stakeholders, capacity building	Regulations, capacity building, public awareness	Development of regulatory framework, capacity building, public awareness	Finalizing processes for regulation development, capacity building, and public awareness

(*continued*)

Table 1.2 (continued)

Biosafety issue	Seychelles	South Africa	Swaziland
Status of development and implementation	Discussion of biotechnology and biosafety issues has only just started in this country to whose economy agriculture contributes only marginally. The main worry is that the country is a net food importer.	Has had a legally binding GMO Act since 1997; also has the institutional framework to administer the act. The country has a number of both public and private laboratories adequately equipped to do GE work. Has more than 110 plant biotech groups, more than 160 plant biotech projects, and more than 150 trials.	Set up a committee within the Environmental Protection Agency to come up with interim legislation on biosafety. Legislation still being developed.
Use of a biosafety system in the regulation of work on or use of GE	Importations of foodstuffs have been handled under the existing food and food standards regulations	Already has a number of GE research work projects and products on the ground, including commercial cultivation of GM horticultural crops, cotton, and maize by smallholder farmers	Has already officially received GM maize under the condition that it has to be milled before distribution to consumers. *Bt* cotton and maize are currently being grown by farmers in parts of South Africa bordering Swaziland, and thus there is fear for possible contamination.
Urgent requirements	Awareness raising, regulations, capacity building	Review of legislation, public awareness and participation	Obtaining stakeholder support, especially from policymakers, as well as regulation development

Source: Based on Mnyulwa and Mugwagwa 2002 but updated through continuous interaction with partners.

Trends in Southern Africa

Currently it is only South Africa that has commercialized GM crops. Both the commercial and small-scale farmers are cultivating these. Below are some figures on the trends of adoption of GM crops in the Makhathini Flats (Kwazulu-Natal Province), the first smallholder farming area to adopt the GM varieties of cotton.

Season	Percentage of farmers cultivating *Bacillus thuringiensis* (*Bt*) cotton
1998/1999	18
1999/2000	60
2000/2001	71

Tanzania	Zambia	Zimbabwe
A national biosafety coordinating committee was set up under the government's Division of the Environment in November 2002. This activity is taking place under the UNEP-GEF project.	Has draft legislation and a national biosafety committee. Limited capacity for risk assessment. Currently in the process of coming up with a national biotechnology strategy.	Has a legally binding biosafety system, which includes a biosafety board and its secretariat, as well as biosafety regulations and guidelines. Has some laboratories, which have the capacity to detect genetically modified organisms (GMOs).
Tanzania has been a port of entry for GM maize provided as food aid to some countries in the region. Consignments were handled under the existing phytosanitary regulations.	An interim committee recommended rejection of GM food aid (July 2002). A case of unapproved trial of GM maize was reported in 1999 (personal communication with Monsanto 2001).	Two field trials were approved in 2001, for Bt cotton and Bt maize. No commercialization has been approved as yet. Assessed applications for importation of GM maize; importation granted with conditions.
Regulations, resource mobilization, public awareness	Enactment of legislation, capacity building, public awareness	Review of current legislation, capacity building, public participation in decision-making processes

GM white maize has been commercialized (2002/03 season) in South Africa, and this will cause a number of smallholder farmers to adopt the cultivation of GM crops.

Overview of GM Use in the SADC Region

The use of biotechnology in the medical sciences is generally well accepted. Its use in agriculture is mixed; for example, South Africa is well into the use of GM crops, while the rest of the SADC nations are still behind. Importation policies are not clear, especially because producers from countries like the United States do not label GMO products.

Public Dialogue, Public Awareness, and Policy Responses

Background

Proponents of GM technologies cite several potential benefits that can accrue to society. These benefits include enhanced taste and quality of foods; nutritional enhancement of foods for chronically malnourished populations; reduced maturation times for crops, leading to labor savings; and enhanced tolerance of biotic and abiotic stresses for crops, leading to reduced dependence on herbicides and pesticides. But these perceived benefits are not uncontroversial.

As a result of the intense debate and controversy surrounding the development and use of GMOs it is important for countries to engage in wide stakeholder dialogues in order to ensure that people are equipped to make informed choices. The public ought to participate even in the development of frameworks for regulation of GM research and development work. The main reasons for public awareness of and participation in the development of national biosafety frameworks (NBFs) are to promote participatory decisionmaking and involve all sectors of the society, to bridge the differences between various parts of society concerning the safe use of living modified organisms (LMOs), to ensure the use of an inclusive process involving all stakeholders, to share a common vision and purpose, to promote improved decisionmaking based on information, and to promote transparency in the decisionmaking process. It is important to note that the development of NBFs goes beyond the creation of a document. It inevitably encompasses wider issues about the role of biotechnology and requires ongoing participation in biosafety processes after regulations have been developed. The process itself calls for commitment and the creation of an appropriate environment to access participatory mechanisms, capacity building, information dissemination, and strategies for involvement of all stakeholders.

Participation in biosafety is prescribed in Article 23 of the Cartagena Protocol on Biosafety (United Nations Environment Program 2002):

Public awareness and participation:
1) Parties to the protocol shall:
   a) Promote and facilitate public awareness, education and participation concerning the safe transfer, handling and use of living modified organisms in relation to the conservation and sustainable use of biological diversity, taking also into account risks to human health. In so doing Parties shall cooperate, as appropriate, with other states and international bodies;

b) Endeavour to ensure that public awareness and education encompass access to information on living modified organisms identified in accordance with this Protocol that may be imported.

The Parties shall, in accordance with their respective laws and regulations, consult the public in the decision making process regarding the living modified organisms and shall make the results of such decisions available to the public, while respecting the confidential information in accordance with article 21.

Participation is crucial in the analysis of the issues, in decisionmaking and strategic planning, in implementation, and in monitoring and evaluation. Stakeholders can be defined as people from government agencies and the private sector, groups or individuals whose lives and interests could be directly or indirectly affected, and bodies, groups, or individuals with particular knowledge that could be called upon.

Public awareness was defined by the participants of a UNEP workshop on risk assessment and risk management held in Namibia in 2002 as a process of providing universal access to information (providing balanced information in terms of pros and cons), enlightening the public, and thereby providing for informed participation. *Public participation* was defined as involving stakeholders (at all levels of society) in decisionmaking processes (giving everyone a chance to express their views) and taking their suggestions into consideration in making a decision. Public awareness and participation are needed for

1. consensus building on issues that affect people directly or indirectly;

2. ensuring implementation of the decision;

3. building transparency and accountability;

4. facilitating informed participation;

5. achieving a better position from which to take action;

6. facilitating inclusiveness;

7. providing balanced information in terms of pros and cons;

8. harmonizing institutions that provide awareness activities;

9. removing bias;

10. building a sense of ownership and collective responsibility;

11. building stakeholder confidence;

12. bridging the knowledge gap;

13. ensuring sustainability;

14. minimizing conflicts;

15. creating a platform for action; and

16. attracting attention and interest.

Status of Public Awareness in the SADC Region

Different countries in the SADC region have sought to promote and facilitate public awareness and participation in the design and implementation of their NBFs. Different tools and approaches have been suggested by various efforts (see United Nations Environment Program 2003a). Participants at a UNEP-GEF Namibia workshop on risk assessment, risk management, public awareness, and public participation for sub-Saharan Africa held in Namibia in 2002 proposed an action plan for enhancing public awareness and participation in the southern African region (see United Nations Environment Program 2003a).

It is the responsibility of each party to determine the combination of the proposed tools suitable for their specific situation. In most countries in the region the lack of biosafety frameworks is partially attributed to these countries' lack of awareness at various levels of the importance of both the technology and the need for biosafety policy. Table 1.3 summarizes the levels of biotechnology awareness in the SADC countries, including the awareness-raising tools and approaches being employed in the different countries.

The Challenges of Public Participation

The public awareness levels shown in Table 1.3, together with the efforts to arrive at such levels, are confounded by many factors, some of which are discussed in this section.

Commercial confidentiality. One of the major challenges of public participation is defining the limits of confidentiality for the provision of information to the

Table 1.3 Levels of biotechnology awareness and public awareness strategies in Southern African Development Community countries, March 2003

Country	Levels of biotech awareness	Strategies used for information dissemination and awareness raising
Angola	Low (assumption)	Little is known about strategies
Botswana	Low overall	Uncoordinated and sporadic activities, mainly announced through newspaper articles and led by scientists and to some extent the consumer movement
Democratic Republic of Congo	Low (assumption)	Little is known about strategies
Lesotho	Low overall	A few sporadic activities, mainly driven by scientists
Malawi	Average among scientists, low among other stakeholders	Discussions in the form of workshops and meetings, mainly coordinated by Bunda College and the National Biosafety Committee. Other tools are mainly sporadic debates and responses via the local press.
Mauritius	Low overall	A few, largely sporadic, activities coordinated by the National Biosafety Committee
Mozambique	Low, even among scientists	Still largely uncoordinated and reactive efforts for coordination through the Africa-Bio and Southern African Regional Biosafety programs
Namibia	Average to low	Some activities coordinated by the National Biotechnology Alliance, the farmers union, and the consumer movement
Seychelles	Low (assumption)	Little is known about strategies
South Africa	Average among the affluent groups but low among smallholder farmers and general consumers	Formal media and informal channels (including Web sites, leaflets, and public debates) sponsored by a number of nongovernmental organizations and companies such as Africa-Bio, Biowatch, SAFeAGE (South African Freeze Alliance on Genetic Engineering), A-Harvest, and Monsanto. Notices of application for trials or release of genetic engineering (GE) products are published in the government gazette to solicit public comments.
Swaziland	Low overall	A few sporadic activities, mainly driven by scientists
Tanzania	Average to low	A few activities, some coordinated by the National Biosafety Committee, some by scientists, and some by the Commission for Science and Technology
Zambia	Average to low among scientists, low among the rest	A few, largely uncoordinated and irregular, activities such as debates and discussions organized by the National Biosafety Committee, the National Farmers Union, and the consumer movement
Zimbabwe	Average among the scientists, low among stakeholders	Advertisements in the government gazette soliciting public comments. A number of organizations engage in information dissemination (e.g., the Biotechnology Trust of Zimbabwe, the Biotech Association of Zimbabwe, the Consumer Council, the Pelum Association, COMMUTECH (the Community Technology Development Trust), the Intermediate Technology Development Group, and the biosafety board, among others. The main channels used include workshops, seminars, debates, information brochures, radio and television discussions, etc.

Source: Based on Mnyulwa and Mugwagwa 2002 but updated through continuous interaction with partners.

public. A statute on access to information might be needed, or the responsibility for deciding what represents confidential information might be given to the national governments in consultation with the companies concerned.

The costs of various levels of participation. These costs need to be planned for and addressed during the planning period. They have to be dealt with in the context of the limited human, infrastructural, and financial resources of most of the countries.

The diversity of the various developing countries' farming systems and other cultural and social factors. This diversity makes it difficult to come up with a common framework for the involvement of stakeholders in the decisionmaking processes.

High science. How does one simplify highly scientific information to facilitate and increase the comprehension of the concepts by the general public, the majority of whom are illiterate? Challenges exist regarding how to effectively communicate science to a public of such a dynamic background as obtains in most of the developing SADC countries, where stakeholders have different priorities to address and have to deal with a language barrier (explaining science in local languages is impossible in most cases). It is noted that dialogue requires honesty, openness, transparency, and inclusiveness, along with mutual respect and an absence of mistrust. The starting point for dialogue should be the premise that the public has valid views that need to be to be voiced and understood, taking into account room for variance. Public participation has to be based on access to information, and it is necessary for national governments to facilitate the packaging of information in a way that meets the stakeholders' needs.

External influences. Many such influences affect decisions taken by developing countries on the commercial use, risk assessment, and risk management issues related to LMOs. Trade in GM crops and products will be subjected to the international agreements signed by the member states. The majority of the developing countries, SADC countries included, are parties to the World Trade Organization (WTO), and thus the protocol is supposed to allow free and equitable trade. Yet the following issues need to be taken into account:

- GMOs require special clearance mechanisms to allow developing countries to make a choice—to accept or reject GMO goods and not be bound by the WTO provisions alone.

- An exporting country is not liable for damage and environmental pollution due to GMOs.

National laws are needed on labeling both the grain and seed and any blended products. Experience so far has shown that the use of GMOs in developing countries is dictated by trading partners such as the European Union.

The murky interface (food aid, politics, science, and regulations). A number of public concerns resulting from the use of modern biotechnology relate to their impact on trade, the environment, and health. Says David Dickson of SciDev.Net: "On closer inspection, the decision by Zimbabwe and Zambia begins to lose some of its apparent naivety. The real fear officials of these countries are said to have explained to the officers of the World Food Program, is not the health danger that these foods are said to cause. Rather it is that if GM maize seed is planted rather than eaten, there could be 'contamination' of local varieties, and this will mean that the agricultural produce of these two countries, including beef fed on the crops, could no longer meet the 'GM free' criteria demanded by European Markets" (http://www.scidev.net/archives/editorial/comment28.html). A study by Environment and Development Activities in Zimbabwe after the 1991/92 drought revealed that about 20 percent of the smallholder farmers from some selected districts of Zimbabwe had retained the yellow maize grain provided as drought relief to use as seed. So the danger that GM maize grain will find its way into the seed system is real.

Most of the developing countries' positions are compromised by those of their trade partners, whether Europe or America. The conflicting positions of the two major trading partners of most southern African countries has greatly influenced the current positions adopted by the various nations.

The United States, one of the major suppliers of food relief, has been commercially growing GM crops for the past 5 or 10 years, and they do not segregate or label these products. The political dimension of the debate over southern African hunger and GM maize is that the United States appears to be using the current famine as a cover to promote acceptance of a technology "enthusiastically embraced by its own corporations, while remaining widely distrusted in Africa" (Dickson 2002). The United States has shown frustration with African critics of its food offer, and has also shown reluctance to provide funds for processing the maize, conditions that have further fueled the political dimension. A statement in early 2002 by one U.S. official that "beggars cannot be choosers" has further haunted the humanitarian effort.

The absence of regulations for monitoring the movement of GM material in most of the affected countries is another problem. Personal communications with some authorities in Zambia have shown that although the trade, food safety, and environmental dimensions have been mentioned, one salient but important dimension has not: that of regulations. The affected parties have feared that lack of a legal framework would frustrate any efforts to ensure monitored and controlled movement of the GM maize once it was released to the population. The situation in Zimbabwe has been different because regulations were in place already, and Malawi (then) was at an advanced stage in the development of its regulatory framework; hence it has been possible for decisions to accept the GM maize to be made.

The situation that has been faced in southern Africa points to the reality that countries have to accept regarding the impact of modern science on society—that it involves a complex of scientific, economic, and political factors that cannot easily be reduced to any single dimension (Dickson 2002).

The Public Awareness Effort in Southern Africa— A SWOT Analysis

Below is a strengths, weaknesses, opportunities, and threats (SWOT) analysis (Table 1.4) of the public biotechnology awareness effort in southern African countries. This analysis is adapted from results of the UNEP-GEF workshop held in Windhoek, Namibia, in November 2002.

Recommendations

Mindful of the situation prevailing in the SADC region with respect to biotechnology, and cognizant of the role that the technology can play in agriculture and food security issues, we recommend that the following needs be addressed.

Development of the Capacity to Make Decisions

One critical issue that emerged from the 2002 debate on food security vis-à-vis the use of GM maize as a food aid was that the majority of countries in the SADC region lacked the regulatory and scientific structures necessary to take decisive steps. During the BTZ's regional consultation on the status of development of biosafety systems in eastern and southern African countries, it emerged as a major sticking point that most countries did not prioritize development of regulatory structures for biosafety, mainly because of the low level of biotechnology research and development activities in their countries. If the lessons drawn from the 2002 GM food aid debate are anything to go by, countries in the region are best advised to put regulatory and scientific monitoring mechanisms in place, because the GM products in the region are not the products of research efforts in the region, but

Table 1.4 Strengths, weaknesses, opportunities, and threats analysis of public awareness and public participation in southern Africa, November 2002

Strengths	High literacy level
	Political will (many countries in the region have signed the Biosafety Protocol)
	Common official language, facilitating information dissemination
	Existing administrative structures
	Information-sharing structures
	Existing human resources (biotech specialists, etc.)
	Relevant legislation and policies
Weaknesses	Limited programs on and capacity for modern biotechnology
	Lack of policies on biotechnology and biosafety
	Ignorance of biotechnology, which impedes the dissemination of information
	Lack of sustainable funding
	Science illiteracy
Opportunities	Existing public awareness and participation programs that can be used to disseminate information, e.g., HIV/AIDS awareness programs
	Decentralized system of governance
	Availability of UNEP-GEF funding
	Existing subregional programs (SADC)
	Innovative financial instruments that could be used to generate additional funds for programs in the form of taxes, levies, and other fees
Threats	Lack of networking among scientists and with other political and civic leaders
	Lack of communication between scientists and other interest groups such as sociologists, politicians, and civil society

Source: United Nations Environment Program 2003b.

rather are products introduced from elsewhere. The scenario is the same as that for products of most other technologies, but the need for regulations remains critical. The GM debate underlined the fact that in a globalized economy the development of regulations is a necessity, not a luxury.

The development of scientific and infrastructural capacity is not an overnight activity. Given the varying levels of capacity and resource endowment in the countries of the region, mechanisms for collaboration and the development of synergistic relationships need to be put in place for countries to be able to pool their resources. Through the SADC and regional as well as national governmental and nongovernmental organizations with activities in the areas of agriculture, the environment, and biotechnology and biosafety, activities can be implemented for the development and strengthening of national and regional capacities that will enable informed decisionmaking on GM products. Arrangements for the transfer of technology and expertise should also be entered into with institutions within the region and beyond that can provide such expertise. Individual countries and the region should place an emphasis on developing their own capacity to do the work so they can become self-sufficient in the long run.

The SADC countries should also be cognizant that genetic engineering is building on the achievements of other accepted and established techniques such as tissue culture, molecular biology, fermentation technology, and so on. Countries need to develop a capacity for these techniques, not necessarily to use them as a foundation for genetic engineering, but to exploit them and assess whether some of the agricultural production constraints can be solved using such technologies. Examples abound from Colombia, India, Kenya, and Zimbabwe, where tissue culture programs have been successfully implemented to provide sufficient quantities of high–health status planting materials for crops such as bananas, yams, cassava, and sweet potatoes.

Identification of Regional Needs and Priorities
For the region and individual countries to realize some of the benefits to be derived from the employment of modern biotechnology techniques, they need not only to develop regulatory and scientific capacity, but also to identify needs and priorities for intervention at national and regional levels. Priorities would include targeting crops or animals for the research efforts, along with traits to be researched (drought tolerance would be an obvious choice) and the human and infrastructural capacity needs of the countries and the region. Genetic engineering technologies invariably need substantial financial investment, and the SADC countries would best be advised to invest in areas in which they have sustainable competitive advantages or in areas that address their priority food security needs.

Creation of an Enabling Environment for Research about or Use of Biotechnology Products
The development and implementation of regulations is one avenue for creating an enabling environment for biotechnology research and development as well as for the use of products of genetic engineering. The SADC countries need to develop appropriate biosafety systems for monitoring and controlling biotechnology activities in them. Given that the region already has three countries with legal biosafety systems, experience-sharing mechanisms can be put in place and employed so countries can learn from each other about the development and use of such systems. Discussion among policymakers needs to be stepped up so as to garner the necessary political will. For example, in Zambia efforts to put policies in place are thwarted not only by lack of funding and scientific expertise, but also by lack of political will. This certainly is the case in most of the countries of the region.

Stakeholders need to develop strategies for ensuring that national governments prioritize policy development and investment in infrastructural and human capacity for biotechnology activities, and at least some measurable capacity for risk assessment and risk management. In a 2001/02 eastern and southern African study on the status of development and implementation of biosafety systems conducted by the BTZ, one of the major findings to emerge was that the source of information most trusted by the lay public was one to which local researchers would have made a contribution. One way to achieve this end is to raise the general level of discourse about biotechnology issues both in the individual countries and at the regional level. With an increased awareness of the potential dangers and benefits of genetic engineering technology, policymakers will be in a better position to see the need to develop the necessary legislative frameworks. Awareness also needs to be raised in the general population of the SADC region because people have a right to know whether they should consume certain products. In addition, transparency and trust need to be developed among the private sector, local researchers, national governments, and all stakeholders in the region with respect to the real hazards or benefits presented by genetic engineering technology.

Harmonization of National and Regional Policies

One major lesson from the food aid debacle is that the countries of the SADC region need to harmonize their legislation in order to facilitate smooth movement and transit of food materials. This harmonization should encompass issues such as standards, risk assessment and risk management procedures, prior informed consent requirements, information and documentation requirements, and other issues. In essence the harmonized policies should facilitate the development of procedures for approval of the use and movement of products in the region.

Conclusion

The SADC countries are at different levels in the development and application of biotechnology as well as systems to govern the use of this technology. This scenario should be exploited to ensure that all countries attain a certain minimum level of technical and regulatory capacity, especially for monitoring the development and use of GM technologies and the products thereof. It is crucial for all the countries in the region to realize that they need each other, especially given the increasingly globalized economy and the fluid nature of national boundaries, as well as the limited capacity to monitor cross-border movement of materials. Adequately equipping the general public, especially farmers, will go a long way toward building self-monitoring and -policing mechanisms that will complement efforts by regulatory authorities to limit the unintended spread of GM products in the environment. An informed society will also influence the national research agenda, thereby ensuring that the constrained research and development resources of countries in the region are used to address priority issues. Little is known about the existing institutional framework within which GMO legislation and regulation are likely to be implemented, especially in rural areas. Several questions therefore remain unanswered. For instance, what roles are played by the national, provincial, and local governments in the various countries? What scientific testing infrastructure exists to implement regulations? What are the existing leadership structures, especially in rural areas? To what extent will uninformed smallholders rely on opinions, information, and advice from village-level leaders in making their choices? What problems and opportunities will result from using the rural governance already in place as a coordinating mechanism for spreading information? What is the degree of transparency and accountability in implementing agencies?

Appendix: Tools for Participation, Consultation, Information, and Education

The following tools have been adapted from United Nations Environment Program (2003b) and from the author's workshop notes.

Tools for Participation and Consultation

There are a number of strategies or approaches that can be used to engender public participation in discussion on biotechnology issues. Some of these are as follows.

Enabling legal frameworks. Laws on public participation or on rights to information facilitate meaningful public involvement in biosafety decisionmaking.

Routine opportunities for public comment. In many countries, applications for regulatory approval are published in a register with opportunities for public comment as a matter of routine. Although this methodology is commonly used in developed countries (for instance, in Canada, the Netherlands, and the United Kingdom), it may be especially useful in developing countries, where there are usually limited resources to facilitate participation.

Multilevel consultations. In some countries, public consultations on different aspects of the biosafety framework have taken place at the national level. For example, consultations were held in Zimbabwe to decide whether to accept GM food aid and, once the decision was made to accept it, how to handle the products.

Independent public inquiries. Independent bodies can be designed to facilitate assessment of the risks and benefits of a technology considering broad public interests. These bodies, if well constituted, can target the particular needs of indigenous groups.

Independent advisory committees. The authority and credibility of such bodies depend heavily on their independence of the government and the way they are constituted, that is, the extent to which they include the views of nonscientists and represent a broad spectrum of stakeholders. These are the tools used by most of the SADC countries, such as Malawi, South Africa, and Zimbabwe. In some cases these are complemented by advertisements in either the government gazettes or the local press soliciting comments from the public.

Ongoing oversight and evaluation. Stakeholder bodies, such as the African Biotechnology Stakeholders' Forum, can be set up to review biosafety procedures on an ongoing basis.

A bottom-up participatory process. Participatory processes facilitated by credible and experienced nongovernmental organizations can help stakeholders at risk of being left out by the government-led consultation processes. Examples include the Citizens Jury facilitated by the Intermediate Technology Development Group in Brazil, India, and Zimbabwe.

These tools can be used in combination to facilitate the all-inclusive participation of stakeholders in the decisionmaking process. The challenges presented earlier in this chapter hinder such effective participation in most developing countries.

Tools for Information and Education

The identification of information gaps through surveys is a good starting point for any awareness and education initiatives. Information collected through these means would help a country's government in the development of a public information campaign using the following tools.

Informal means of disseminating information. Web sites, leaflets, advertisements, and telephone help lines can be used to explain biosafety processes and how stakeholders can be involved in information dissemination. These can even be translated into local languages. The BTZ has been using some of these methodologies in disseminating information to the rural poor.

The established media. Newspapers, radio, and television provide useful routes for informing the public about biotechnology and biosafety regulations. These can be used to educate or inform the public about GMOs. Advertisements can also be used to get feedback on proposed releases of GM products.

References

Dickson, D. 2002. African hunger and GM maize. http://www.scidev.net/archives/editorial.

James, C. 2001. *Preview: Global review of commercialized transgenic crops, 2001.* ISAAA Briefs no. 25. Ithaca, NY: International Service for the Acquisition of Agribiotechnology Applications (ISAAA).

Mnyulwa, D., and J. T. Mugwagwa. 2002. *Agricultural research needs for southern African countries: Towards a regional initiative on need-driven agricultural biotechnology.* Harare, Zimbabwe: Biotechnology Trust of Zimbabwe.

Persley, G. J., and J. N. Siedow. 1999. *Applications of biotechnology to crops: Benefits and risks.* CAST Issue Paper 12. Ames, IA, USA: Council for Agricultural Science and Technology.

Sasson, A. 1993. *Biotechnologies in developing countries: Present and future.* Vol. 1, *Regional and national survey.* Paris: United Nations Educational, Scientific, and Cultural Organization Publishing.

United Nations Environment Program (UNEP). 2002. Cartagena protocol on biosafety. UNEP Web site (http://www.unep.org).

———. 2003a. DFID/UNEP [Department for International Development / United Nations Environment Program] study on public awareness and participation. UNEP Web site (http://www.unep.org).

———. 2003b. Report of the Regional Workshop on Risk Assessment, Risk Management and Public Participation, Windhoek, Namibia, November. UNEP Web site (http://www.unep.org).

Chapter 2

Consensus-Building Processes in Society and Genetically Modified Organisms: The Concept and Practice of Multistakeholder Processes

David Matz and Michele Ferenz

This chapter begins by outlining key conceptual issues in multistakeholder processes. Three examples of such processes from across the globe are then presented: first, an electronic multistakeholder dialogue from India; second, scenario workshops from Denmark; and third, a rights-based approach from the World Commission on Dams. The three examples have been selected to illustrate key issues outlined in the three conceptual pieces. Although the examples do not focus on biotechnology, and although only one of them is from a specific developing country (India), together they help build understanding of the kinds of conceptual and practical issues that must be addressed in multistakeholder processes. It is also important to recognize that the various attempts to raise awareness and build consensus on biotechnology in developing countries have not been explicitly conceived or implemented as multistakeholder processes in the sense that they have not taken full account of the central challenges facing such processes. These challenges are outlined here, along with the most promising approaches to addressing them.

The Concept of Multistakeholder Processes

Whether in dialogues or in partnerships, a multistakeholder approach is fundamentally about negotiation between different sectoral and societal interests. Conventional wisdom regarding negotiation sees the activity as inherently defensive, and often manipulative. It is often assumed that adversarial position taking and concession trading is the only way for each party to achieve a solution that meets his or her minimum demands. Parties to a negotiation, it is believed, artificially inflate demands and dissemble to avoid appearing "weak," a condition that would be immediately exploited by those on the other side.

Yet this adversarial approach usually produces only "lowest-common-denominator" project, program, and policy outcomes. These outcomes are almost never sustainable over the long term, environmentally or in any other way. If people feel coerced or cheated in some way during a negotiation, they will fail to live up to the agreement. What is more, when people feel excluded from decision-making, when they are not given "a voice at the table," they will not identify with the directives agreed to and will ignore or even boycott them.

Even the so-called winners in a negotiation conceived of as a strategic cat-and-mouse game often could have done much better with an "integrative bargaining" approach. Such an approach rejects the logic of aggressive destabilization and undercutting of the "opponent." Instead, it recognizes that parties in negotiation almost always have both competing and complementary or compatible interests. The challenge then becomes to structure the negotiations such that these common interests are allowed to emerge so that they may serve as the basis for a mutually satisfactory resolution. In short, the negotiation becomes a joint discovery and problem-solving exercise that typically moves through the following stages.

1. *Information gathering and exchange.* The key is to focus the deliberations on needs and interests and the reasons underlying the positions typically put forth as demands in negotiations. An example highlighting the difference between positions and interests can be drawn from the Camp David talks between Israel and Egypt, which bogged down over the issue of control of the Sinai Desert (the position "control is ours"). When it became apparent that Israel wanted to retain control for "security reasons" (Israel's interest), whereas Egypt was primarily interested in restoring its "sovereignty" as a nation (Egypt's interest), the stalemate could be broken. Based on this revelation, an arrangement was forged that addressed both interests, though through different means than the ones demanded by the respective parties (because, of course, it was impossible to simultaneously give control of the territory to both disputants).

2. *Invention of possible options.* Parties should be given the opportunity to put forth proposals that meet their needs as well as those of other stakeholders. The best way to elicit creative thinking in this phase is to assure participants that they will not be bound by any suggestions they make at this stage, which separates

inventing from committing. This is meant to be a brainstorming phase during which people can bounce ideas off each other, and can build on others' proposals or modify them to make them more acceptable ("reality testing").

3. *Packaging.* Negotiations are rarely about one issue alone; a conflict can be disaggregated into multiple elements, and the parties are likely to have differential priorities and preferences that can be capitalized on to maximize joint gains by trading across issues. For example, if X is very important to me and Y less so, and for you the preference ranking is the reverse, we will likely be able to find a settlement whereby I will get more of X and you will get more of Y. To ascertain such preference rankings (because often they are not clear even to the negotiator unless he or she is faced with making choices) and to engineer the trading game, the parties should consider several different packages of options and jointly piece together the one that is the best fit for as many parties as possible.

4. *Finding mutually acceptable criteria for dividing joint gains.* Inevitably a negotiation hits a point at which trades are no longer possible. It then becomes what is often called a "zero-sum game," meaning that some parties will be able to extract a better outcome for themselves than will others. As implied earlier, many negotiations start with this dynamic, and the purpose of phases 1–3 is to delay it long enough for creative solutions to emerge and for positive relationships to solidify between the parties. In order not to undo all that hard work, it is important at this stage to jointly establish criteria that will guide the division of the gains created. Such criteria may include efficiency and equity considerations or make reference to ethical principles, community practice, or legal precedent. Such criteria not only ensure that the process of division will not break down into a mere show of force; they can also serve as points of orientation in the next negotiation among the same parties (because of professional affiliations, community ties, and so on, parties typically find themselves reunited in different negotiating fora again and again).

5. *Including contingency plans and monitoring provisions.* Often the most difficult phase begins once the agreement is signed. Not only are resource constraints a common problem that inhibits implementation; agreements are often based on assumptions that turn out to be wrong. Because it is impossible to predict the future, uncertainty is an inherent factor to contend with, and this problem is especially acute when dealing with science-intensive environmental issues. It is important to account for uncertainty and render the agreement robust in the face of this uncertainty by building into the accord itself contingency plans (if A happens, we agree to do X; if B happens, we agree to do Y) as well as provisions for ongoing consultation and dispute resolution mechanisms. To ensure that a group of stakeholders is able to move through these various phases, the services of a professional nonpartisan facilitator or mediator may be needed. Facilitation is the nonintrusive

management of an exchange of views between parties; it ensures that all parties are heard and minimizes misunderstandings. Mediation is "assisted negotiation," the shepherding of the parties through a structured process that aims to achieve an agreement or plan of action. Third-party intervention is especially desirable when the issues at stake are multifaceted and complex or when relations between the parties are characterized by hostility at the outset. Indeed the difficult task of hearing out opposing interests, lessening fears, and opening minds is a key purpose of multistakeholder efforts and a precondition for multiparty on-the-ground execution of joint action plans. As a publication of the Mining, Minerals, and Sustainable Development Project (described later) asserted, "One of the Project's main outcomes will be the set of relationships it is building through this process and their capacity to continue, and perhaps implement, a change agenda in the future" (IIED/WBCSD 2001).

The Shift to Participatory Planning and Multistakeholder Dialogues

The recognition that top-down approaches often do not produce the desired results has led to what might be characterized as a radical shift in development policy over the past decade. While some key development-related institutions (especially the international financial institutions, such as the International Monetary Fund and the World Trade Organization) are still largely closed to perceived outsiders, many government organizations have, to varying degrees, opened their doors to civil society.

Indeed the years since the 1992 United Nations Conference on Environment and Development (UNCED) have seen a virtual explosion of experimentation with multistakeholder approaches, both at the national level and increasingly at the international level. These usually take one of two forms:

Site-specific approaches. An example would be the placement in an ecologically sensitive area of a polluting coal-fired power plant considered vital to the economic development of the region. Here representatives of affected government, business, environmental, and community interests would together work out a construction, mitigation, or compensation package. At this project level, participatory planning is intended to ensure that intended beneficiaries as well as those potentially negatively affected by a project have a say in the conceptualization and implementation of a particular economic development scheme or planning measure. Where appropriate, so-called "local knowledge" should be heeded to tailor generic program blueprints to specific contexts and circumstances and to disrupt as little as possible the social, economic, and ecological fabric of communities that are to be the project hosts.

Policy-focused approaches. An example would be working out guidelines for and elements of a national energy policy, elaborating rules governing hazardous waste disposal, or devising recommendations for future large hydrological projects, as was done by the World Commission on Dams recently. Here consultations take on various forms. In the United States, a practice that has come to be known as "negotiated rulemaking"—the involvement of stakeholders in the crafting of administrative provisions that serve to interpret and enforce legislation—has become quite common in the environmental arena. Those efforts are led and brokered by the responsible executive authorities, such as the Environmental Protection Agency. Sometimes stakeholders themselves, alone or in conjunction with others, launch a multistakeholder initiative. One example is the Mining, Minerals and Sustainable Development Project, a two-year effort of participatory analysis of the sector managed through the International Institute for Environment and Development (IIED), the World Business Council for Sustainable Development, and a global network of regional partners, which canvassed stakeholders from the world's biggest mining companies to some indigenous communities. Through commissioned papers, thematic workshops, and interviews the project has generated a substantial database of information, some of which was synthesized in the final report issued in 2002 (IIED 2002). More and more frequently, different policy enterprises of this sort are loosely grouped under the umbrella term "multistakeholder dialogues" (MSDs).

The important point is that MSDs—whether organized by nongovernmental organizations on a one-time-only basis or structured as ongoing exchanges organized by a country or a multinational organization—bring nongovernmental actors into the conversation. While multilateral policymaking organizations—such as the United Nations, the Organization for Economic Cooperation and Development, and the World Bank—remain entities to which only countries can apply for membership, these institutions are increasingly finding that they must incorporate the views and inputs of nongovernmental interests in order for their work to be seen as legitimate and to gain access to the relevant knowledge and skills required for complex problem solving. In a sense, this is the culmination on a global scale of a trend that took hold as far back as twenty years ago in planning efforts at the local, regional, and national levels in the United States, Canada, and Europe and is fast spreading to other parts of the world.

Lessons Learned from Multistakeholder Initiatives to Date

Scholarship assessing the proliferating multistakeholder initiatives is in its early stages. Nonetheless, it appears fair to conclude that experiences to date have highlighted four particular challenges in the organization of multistakeholder efforts for sustainable development: We will deal with each of these in turn.

1. Ensuring that all the relevant parties are involved in negotiations. Carlson (1999) defines stakeholders as "key individuals, groups, and organizations that have an interest in the issue at hand. They may be responsible for seeing a problem resolved or a decision made, they may be affected by a problem or decision, or they may have the power to thwart a solution or decision." The values or interests they represent often categorize stakeholders. Some institutions divide stakeholders into three groups—government, business, and civil society. However, more fine-grained distinctions among stakeholders have sometimes been made, especially in UN proceedings since the 1992 Earth Summit identified nine major groups—women, children and youth, indigenous people, nongovernmental organizations (NGOs), local authorities, workers and trade unions, business and industry, scientific and technological communities, and farmers (a chapter is dedicated to each of these in line with its openly participatory vision in Agenda 21 [UNDESA 1997], a comprehensive plan for safeguarding the environment that was adopted by the countries participating in the seminal UNCED). The World Commission on Dams created an advisory forum to act as a sounding board for its commissioners, which included 68 stakeholder organizations. After a closer examination of the large-dams policy arena, the World Commission on Dams distributed representation on the forum across ten stakeholder categories, including private sector firms, river basin authorities, utilities, multilateral agencies, bilateral agencies and export credit guarantee agencies, government agencies, international associations, affected people's groups, NGOs, and research institutes.

Involving such varied constituencies requires that each be sufficiently organized to speak with something approaching a unified voice. Completing internal negotiations in which each group irons out its own differences before the larger dialogue begins may be very difficult. The negotiation process must therefore require transparency and viable modes of access for all interested groups (depending on the situation, Web-based communication may be an appropriate tool). It must also allow for repeated rounds of consultation and be structured as a continuing sequence of inside-outside negotiation. Such a structure promotes ongoing feedback and forestalls the tendency of negotiators to lock into one position before hearing the others. It also ensures that the representatives are accountable to their constituencies and do not stray from their wishes in a way that would imperil the wider acceptability of an agreement.

A technique called "conflict assessment" helps ensure that the right parties are involved in the negotiations (see Figure 2.1). As part of such an assessment, an

Figure 2.1 How to conduct a conflict assessment

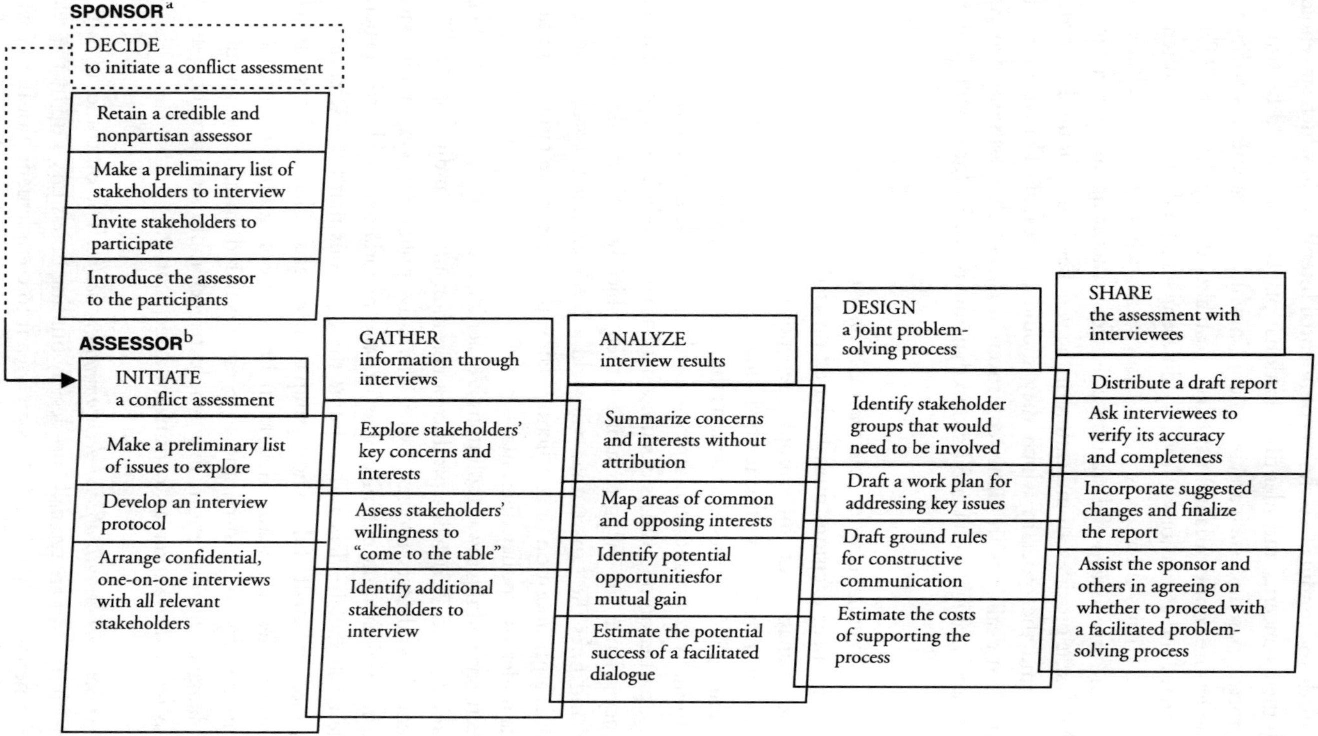

SPONSOR[a]

- DECIDE to initiate a conflict assessment
  - Retain a credible and nonpartisan assessor
  - Make a preliminary list of stakeholders to interview
  - Invite stakeholders to participate
  - Introduce the assessor to the participants

ASSESSOR[b]

- INITIATE a conflict assessment
  - Make a preliminary list of issues to explore
  - Develop an interview protocol
  - Arrange confidential, one-on-one interviews with all relevant stakeholders

- GATHER information through interviews
  - Explore stakeholders' key concerns and interests
  - Assess stakeholders' willingness to "come to the table"
  - Identify additional stakeholders to interview

- ANALYZE interview results
  - Summarize concerns and interests without attribution
  - Map areas of common and opposing interests
  - Identify potential opportunities for mutual gain
  - Estimate the potential success of a facilitated dialogue

- DESIGN a joint problem-solving process
  - Identify stakeholder groups that would need to be involved
  - Draft a work plan for addressing key issues
  - Draft ground rules for constructive communication
  - Estimate the costs of supporting the process

- SHARE the assessment with interviewees
  - Distribute a draft report
  - Ask interviewees to verify its accuracy and completeness
  - Incorporate suggested changes and finalize the report
  - Assist the sponsor and others in agreeing on whether to proceed with a facilitated problem-solving process

Source: Consensus Building Institute 2001.

[a] A sponsor is any individual or organization interested in assessing the feasibility of a facilitated dialogue.
[b] An assessor must be neutral, impartial, and experienced in dispute resolution.

impartial mediator conducts a series of confidential interviews in which stakeholders clarify their concerns and identify additional players that should be brought into the process. Based on such an assessment (in which no statement is attributed by name to ensure confidentiality) a mediator can also identify the degree of overlap of the views and aspirations offered by different stakeholders (which often are closer than the parties themselves realize). Such an analysis of potential areas of agreement can serve as a useful starting point for structuring an agenda for the ensuing MSDs; it provides an indication of the way key issues should be worded and framed, and the order in which they should be treated. This is especially important when dealing with highly controversial issues, when tensions between the groups can run high and a good group dynamic is crucial for moving toward consensus.

2. Getting accurate scientific and technical information on the table. Environmental management decisions must be based on credible scientific and technical input. Water management, for example, depends on matters such as the hydrological and ecological effects of watershed modification, supply and demand forecasts for a multiplicity of uses, and actions that can help maintain and enhance the resource. In many court and legislative proceedings, as well as in many larger policy debates, parties on opposing sides use what has come to be known pejoratively as "advocacy science" in trying to support their objectives. Each side frames the questions and hires the experts that will yield a predetermined "correct" answer. The result is a juxtaposition of conflicting claims that exacerbate rather than help resolve the underlying policy dilemma.

Collectively working toward solutions is easier if a process of "joint fact-finding" (see Figures 2.2 and 2.3) helps produce a common understanding of the likely effects, benefits, and costs associated with alternative policy options. In joint factfinding a neutral facilitator typically assists the negotiators to identify experts acceptable to all stakeholders and to frame the questions that these scientists are commissioned on behalf of the whole group to investigate. Their findings can help reduce uncertainties and factual disagreements, set priorities for action that may differ from country to country, and help establish "red lines," or thresholds of resource damage and depletion, that would trigger more stringent obligations (known, as referred to earlier, as "contingent agreements").

3. Promoting links with official decisionmaking bodies. The outcomes of multistakeholder initiatives are typically not legally binding unless taken up by the relevant governmental authorities. MSDs are meant to complement, not in any way to supplant, the legitimately constituted decisionmaking channels (nor are they

CONSENSUS-BUILDING   45

Figure 2.2 Key steps in the joint fact-finding process

UNDERSTAND the interests and issues at stake STEP 1	DETERMINE whether JFF is appropriate STEP 2	SCOPE the JFF process STEP 3	DEFINE the most appropriate methods of analysis STEP 4	EVALUATE the results of JFF STEP 5	COMMUNICATE the results of JFF process STEP 6
Initiate discussions to enhance understanding of the issues	Determine if a fair fact-finding process is possible	Determine which stakeholders need to be involved and whether they will agree	Assess the information that exists and identify information gaps	Agree on methods for dealing with conflicting interpretations	Jointly review and discuss final drafts of reports and studies
Meet with constituencies to (a) develop a set of issues to be addressed and (b) frame a mission statement	Identify strategies for responding to significant power imbalances	Formulate questions to be answered and issues to be addressed	Reframe general questions as specific questions to put to experts of various kinds	Construct a sensitivity analysis to examine the overall significance of scientific disagreements	Determine whether and how JFF results have (or have not) answered the most important questions
	Identify funding for JFF	Formulate ways of answering the questions	Identify various methods of information gathering and analysis and highlight the benefits and disadvantages of each	Complete the data gathering and analysis	Determine if further JFF is necessary
		Establish and agree to procedural and conversational ground rules	Determine the values and shortcomings of various synthesis or decisionmaking tools	Clarify the remaining uncertainties and possible contingent responses	Integrate findings into recommendations
		Determine the role of experts vs. nonexperts and of policymakers vs. other stakeholders	Complete the preliminary data gathering and analysis		Communicate results to various constituencies and policymakers
		Agree on mechanisms to counter severe disparities in expertise			

Source: Consensus Building Institute 2002.

Figure 2.3 The consensus-building process and the role of joint fact-finding

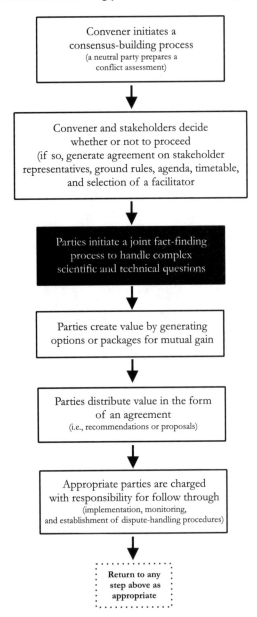

Source: Consensus Building Institute 2002.

intended to serve as lobbying sessions). The style of these dialogues often differs from that of traditionally more rule-bound and hierarchically structured diplomatic negotiations or administrative proceedings. The best results are typically achieved when relative informality characterizes the deliberations; an open, free-flowing dialogue, preferably facilitated by a skillful chair who enforces agreed-upon ground rules to ensure equity and civility, allows for creative problem solving and (often) allows consensus positions to emerge.

Consensus is achieved when almost all participants agree that they can "live with" a proposed "package" after every effort has been made to address the interests of the participants. In practice, while MSDs seek unanimity, most reach a point at which an overwhelming majority agrees, but one or two have more to gain by dissenting. If the group discovers, after probing the concerns of the holdouts, that nothing more can be done to meet the interests of those who do not agree, they conclude their efforts (Susskind 1999). It should be remembered, however, that reaching consensus is not an absolute requirement in every case. When MSD designers are hoping to build new relationships, generate a new way of framing a seemingly intractable problem, or pass along new information, a non-consensus-seeking process may be most appropriate. The aim then becomes to generate "some good ideas" or the group's "best advice."

Still, the judicious use of outputs from an MSD—whether consensus-based or not—is crucial. Parallel processes engaging key stakeholders in the generation and evaluation of options and the formation of partnerships in furtherance of policy objectives should not be held in a vacuum. Constraints on enforcement are, of course, not limited to civil society processes. Nonetheless, the ad hoc nature of multistakeholder efforts makes it important to pay particular heed to the possible transformation of informal understandings into binding commitments or into recommendations that will be useful to, and therefore taken seriously by, the designated governmental decisionmakers. Ground rules for engagement and rules for channeling outcomes into official deliberations must be clarified from the beginning. Policy dialogues and alliances are painstaking endeavors, and civil society representatives will become disillusioned and distracted if their efforts are not given due consideration. Along with the responsibility that is increasingly assigned to civil society for realizing the transition to sustainable development should go the right to claim respect and recognition for the expertise and experience contributed.

Of course in order for the civil society recommendations to be taken seriously, they must be credible and well founded. This in turn requires that dialogue delegates be adequately prepared for the deliberations. The uneven quality of participant contributions is a complaint that commonly arises with respect to MSDs. Capacity constraints are one of the major obstacles to effective participation. This is a problem particularly when stakeholders with vastly different resource endowments come to the table together. Again, the responsibility for evening out the playing field as much as possible falls to the mediator or facilitator, who can identify gaps in knowledge and coordination abilities faced by individual stakeholder groups and help overcome these, all the while being transparent with all parties about the process principles and guidelines to be followed to prepare participants for a useful exchange.

On the governmental side, appropriate national and local legislation matching the intent of an agreement reached can be crafted only when an adequate regulatory apparatus is in place. In many parts of the world where environmental problems are most acute, few people are available who have the background to engage in the monitoring, modeling, and analysis of technical and regulatory options. Transferring the requisite skills and housing such expertise in local institutions—governmental and nongovernmental—that are strong enough to muster adequate resources and autonomy from vested interests are priority concerns.

4. Establishing fairness and efficiency as criteria for evaluation of multistakeholder processes. I refer here principally to procedural fairness (or "due process" as it is known in the legal realm), which is most often measured in terms of stakeholder perceptions. It requires transparency and predictability of the proceedings as well as the preparatory stages that lead to them and the implementation stage that follows. It is paramount that all participants be given equal access to key information and equal opportunity to air their views. Efficiency, on the other hand, is a measure of the quality of the outcome. Here the key question is whether all plausible options were explored and all possible opportunities exploited. If potential "joint gains" are left "on the table"—that is, if information valuable to some stakeholders is left unstated by others, if partnerships that could have been fail to form, or if consensus eludes the group despite the compatibility of different interests—benefits were not fully captured.

A Checklist of Questions to Be Answered about How to Make Policy Concerning GMOs

Following is a list of questions that need to be answered concerning the process to be used in making policy under conditions characterized by multidimensionality and complexity such as those involving biotechnology and genetically modified organisms (GMOs). It roughly parallels issues to be addressed in the phases of building agreement listed in Figure 2.4 and is meant to give an overview of what policymakers need to consider. On the pages following this list are three brief examples that explore some answers to some of these questions.

CONSENSUS-BUILDING   49

Figure 2.4   Phases of building agreement

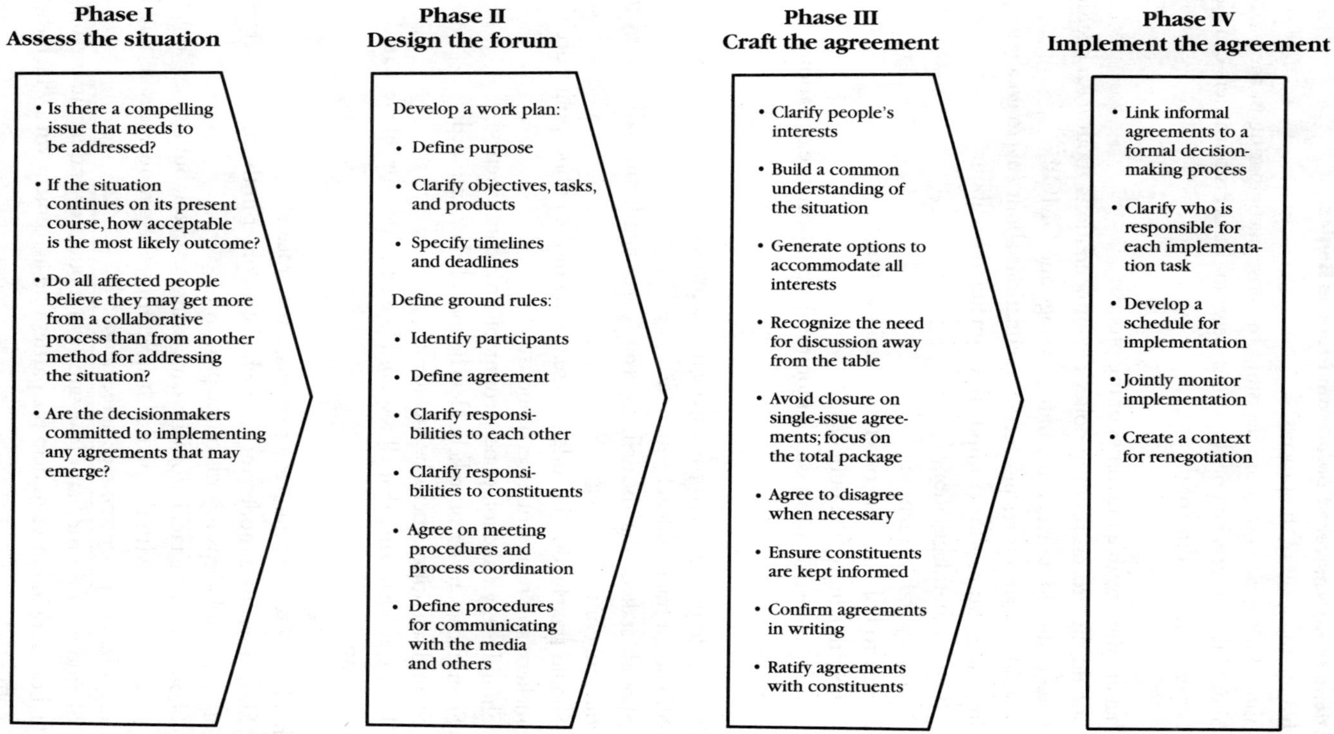

**Phase I
Assess the situation**

- Is there a compelling issue that needs to be addressed?
- If the situation continues on its present course, how acceptable is the most likely outcome?
- Do all affected people believe they may get more from a collaborative process than from another method for addressing the situation?
- Are the decisionmakers committed to implementing any agreements that may emerge?

**Phase II
Design the forum**

Develop a work plan:
- Define purpose
- Clarify objectives, tasks, and products
- Specify timelines and deadlines

Define ground rules:
- Identify participants
- Define agreement
- Clarify responsibilities to each other
- Clarify responsibilities to constituents
- Agree on meeting procedures and process coordination
- Define procedures for communicating with the media and others

**Phase III
Craft the agreement**

- Clarify people's interests
- Build a common understanding of the situation
- Generate options to accommodate all interests
- Recognize the need for discussion away from the table
- Avoid closure on single-issue agreements; focus on the total package
- Agree to disagree when necessary
- Ensure constituents are kept informed
- Confirm agreements in writing
- Ratify agreements with constituents

**Phase IV
Implement the agreement**

- Link informal agreements to a formal decision-making process
- Clarify who is responsible for each implementation task
- Develop a schedule for implementation
- Jointly monitor implementation
- Create a context for renegotiation

Source: Montana Consensus Council 1998.

Questions to Be Answered *before* the Process Begins
- What are the goals of the process?
  - Should the process result in decisions by those participating in the process?
  - Should the process be one for airing views so the decisionmakers can gain a better understanding of the issues?

- What are the possible outcomes of the process?
  - Should the process result in policy recommendations about *what* decisionmakers should do regarding biotechnology and GMOs?
  - Should the process result in recommendations about *how* to go about implementation—for example, recommendations about
    - how to draft legislation,
    - how to draft regulations,
    - how to hold further conferences and meetings,
    - how to educate the public, and
    - how to develop processes to monitor the performance of various players?

- Who should be invited to participate in the process?
  - Should all stakeholders be invited?
  - Should "stakeholder" be defined as any party significantly affected by the outcome of the process?
  - Should stakeholders include representatives of the public, policy decisionmakers, and representatives of industry?
  - Should every stakeholder be accountable to a constituency?
  - Should any one set of stakeholders be included or excluded?
  - Should scientific experts be included?
  - How can we be sure that all responsible scientific points of view are presented?

- Should we use a neutral party to manage the process?
  - Should we use a moderator (one who keeps order in the process, sets the agenda, and keeps records of the process)?
  - Should we use a facilitator (the same as a moderator, but also explores issues in some depth with the parties, helps clarify where differences lie, and helps organize the process to seek agreements)?
  - Should we use a mediator (the same as a moderator and facilitator, but also takes more initiative to help the parties find agreement with which they are comfortable)?

- How does the process selected relate to the larger public dialogue on the subject?
  - What is the role of the media in educating the public about issues and recommendations?

- What kind and size of staff is needed to make the process effective and efficient?

- What level and source of funding are necessary to make the process possible?

- What resources should be planned for (e.g., budgeted for) in advance so that follow-up will be possible after the process has been completed?

Questions to Be Answered *during* the Process
- What rules of decisionmaking will be used?
  - Is unanimity required for any decision?
  - Is consensus (lack of any strong objector) sufficient? (Consensus suggests agreement among all or many of the participants, or at least a willingness by some to go along with the final recommendations.)
  - Are dissenting views to be part of the final report?

- How can all parties be given an opportunity to present their viewpoints to all participants?
  - How can we give speakers a feeling of being heard?
  - How can we give listeners a feeling that they understand what they are hearing?
  - How can we encourage candor in presentations rather than posturing or mechanical restatements of what everyone expects to hear?
  - How can presentations be "translated" across disciplinary and cultural barriers?
  - How can we manage difference inside each stakeholder group?

- How can we frame issues and questions so they can be answered to the extent possible?

- How can we manage scientific information?
  - How can we decide who is an expert?
  - How can we know which questions are predominantly ones of science and which are predominantly ones of politics?
  - How can we ensure that laypeople are comfortable with and knowledgeable about scientific language and judgments?

- How can we help decisionmakers and the public deal with differences among scientific experts?
- How can we help the public and decisionmakers deal with different predictions of the future as different experts express them?

• How can we frame areas of possible consensus or agreement (if these are the goals of the process)?

• How can we hold stakeholders accountable to their constituencies?
  - How can we ensure that representatives have the backing of their constituencies?
  - How can we ensure that representatives can deliver what they agree to?

Questions to Be Answered *after* the Process Ends
• How can we monitor decisions or obligations undertaken during the process?

Examples That Explore Answers to Some of the Questions

India: An Electronic MSD

The following example of a dialogue has been adapted from Scoones and Thompson (2003).

In 2002 a report titled "Prajateerpu: A Citizens Jury / Scenario Workshop on Food and Farming Futures for Andhra Pradesh" was published (Pimbert and Wakeford 2002). The workshop it described had been devised to enable those people most affected by the "Vision 2020" for food and farming in Andhra Pradesh, India—smallholder and marginal farmers—to comment on the development strategy of the state and to shape a vision of their own. The release of the report sparked an international debate over the use of participatory approaches to inform and influence policy from below. Strong views were expressed, and questions were raised about citizen engagement in policymaking processes, the trustworthiness of participatory "verdicts" and the implications that could be drawn from them, ways to increase accountability and transparency in policymaking, and other issues. The E-Forum on Participatory Processes for Policy Change was established and moderated by two researchers at the International Institute for Environment and Development (IIED) in response to this debate. The forum was designed to create a constructive dialogue around certain key issues. This "forum on a forum" sought to draw attention to the important methodological, conceptual, and substantive lessons emerging from the citizens' jury and scenario workshop experiment and encouraged all interested parties to contribute ideas and opinions on key issues arising from the Prajateerpu ("people's verdict") experience.

All those involved in the debate through informal e-mail and other means were invited to participate at the outset. This included the Prajateerpu partners in Andhra Pradesh, the directors and staff of Institute of Development Studies and the IIED, NGO and donor personnel, academics, and other interested observers. Many made contributions. The e-forum ran over 40 days in August and September 2002.

The e-forum was organized around four issues: (1) evidence, (2) representation, (3) engagement, and (4) accountability. These open-ended but generic issues were chosen to allow those not directly involved in the Prajateerpu exercise or familiar with Andhra Pradesh to share their knowledge and insights. A Web site was created to make all the contributions available to those interested. Clear principles of engagement were also set out at the beginning of the process to assure contributors that the moderators would not seek to impose their points of view in the process. A wide range of views were expressed in the forum on each of the issues, and yet in several areas some consensus emerged.

Issues of evidence. Nearly every participant in the e-forum agreed that the Prajateerpu exercise had been a significant effort to develop and extend methodologies for popular participation in policymaking. On the issue of evidence, some commentators believed strongly in a conventional positivist view of knowledge and truth. But the majority of the commentators took a more reflective view of the issue, arguing that all knowledge is necessarily situated and constructed, and that no simple truth can come out of highly contested, complex, and uncertain deliberations about future scenarios of the kind that the Prajateerpu participants had considered.

Several commentators expressed their disappointment that the Prajateerpu exercise (or at least the report) did not seem to capture the range of dispute and debate and the nuances of deliberation among the participants. Others remarked that the commentary of the authors added a layer of interpretation to the participants' statements. They thus raised the question of how facilitators can avoid accusations of partiality and manipulation of results. As these sorts of exercises increasingly come to be used to influence policy, it will be important to address this question, or those who do not like what they are hearing will discredit more deliberative and inclusive engagement.

An underlying theme of many of the contributions was the related question of the politics of methodology. Many commentators agreed that concerns over methodology have been used by those in power to discredit those who challenge a dominant discourse, as was certainly evident in the controversy over Prajateerpu. With a focus of the debate on issues of "quality" defined in narrow, positivist terms, those who objected to the results of the workshop were able to reframe the discussion and divert attention from more pertinent issues. The contributors to the e-forum by and large rejected this position and argued for a more plural and open approach with a wider view of acceptable criteria for evaluating "evidence" and assessing results. Many contributors emphasized the importance of plural perspectives, open debate, and diversity of views. Open deliberations rarely result in neat consensus, let alone a jury-style verdict. Thus many participants argued for more open-ended outcomes than those allowed for in the Prajateerpu exercise.

Issues of representation. Every development organization today seems to need "the poor" to speak in support of their policy positions to give them legitimacy and credibility. Much commentary in the e-forum dwelt on the representativeness of the jurors and the scenarios used as a focus for the deliberations. Many of the contributors acknowledged that representativeness is a contested and loaded term. Several contributors remarked that the Prajateerpu "citizens' jury" was not strictly a jury. The jurors, made up of poor people, mostly women, who were reliant mainly on farming and came largely from a Dalit caste background, had been selected not randomly, but purposively. They were intended to "represent" not society at large, but rather a particular marginalized group with a particular set of interests and livelihood constraints.

Much e-forum commentary also dwelt on the "representativeness" of the scenarios used to inform the Prajateerpu jury's deliberations. Some viewed these as biased, and therefore as creating a "self-fulfilling prophesy." The range of scenarios presented to the farmer-jurists may have limited the debate. Some participants called attention to ongoing research in Andhra Pradesh that highlights a greater complexity of livelihood pathways than was captured in the three scenarios used in Prajateerpu. Perhaps a more interesting route would have been to focus on the trade-offs between scenarios, explore the gap between polarized positions, and avoid the perhaps artificial "verdict."

Issues of engagement. The Prajateerpu event had been only one part of a longer process of policy engagement and debate, the moderators reminded us. Critiques of the Vision 2020 approach adopted in Andhra Pradesh did not start and will not end with Prajateerpu. But to develop an alternative vision for a sustainable rural future, much more work will have to be done beyond simply rejecting Vision 2020 as the farmer-jurists did. Processes of influencing policy outcomes are a critical complement to any deliberative forum or event. How do we locate citizens' juries, panels, or scenario workshops in broader policy processes? The e-forum contributors discussed different alternatives both implicitly and explicitly.

Issues of accountability. To what extent do deliberative processes, such as that used in Prajateerpu, offer opportunities for holding the powerful to account? One of the specific aims of the jury process was to hold the government of Andhra Pradesh and its donors to account, allowing the "beneficiaries" to question their motives and strategies. Follow-up meetings with Andhra Pradesh and U.K. government officials were clearly designed toward this end. The commentary contributed by the Department for International Development (DFID)–India to the e-forum in fact revealed that the process has encouraged reflection within DFID on its approach in Andhra Pradesh, and indicated some success in this regard. But are complex, necessarily expensive, high-profile events like that in Prajateerpu the model for improving accountability? Or are other routes, such as more informal lobbying or the normal channels of representative democracy, likely to be more effective?

Much of the discussion surrounding the Prajateerpu results has been focused on DFID and the U.K. government rather than on the Andhra Pradesh government. Inadvertently the Prajateerpu exercise has raised some important questions about the accountability of aid donors. Is it acceptable for foreign donors to say that their support is granted to an elected government that is responsible to its electorate as to how the money is spent? Participants in the Prajateerpu exercise clearly did not think so. Though this issue was not explored in depth in the e-forum, it will be raised again.

Despite differences of opinion and interest in issues, the e-forum showed much more common ground than first appeared. The insights contributors offered demonstrated that the practical, the political, and the process are all intertwined, and that simple responses based on narrow framings or limited methodological viewpoints are insufficient. The debate the Prajateerpu experience ignited also revealed a number of significant issues regarding the people-centered approaches and processes that can be used to influence policy from below, which were highlighted in the many constructive offerings made to the e-forum. Few issues were resolved, however, and most will require further deliberation. In the future this debate will occur in a range of fora and among a variety of networks. The e-forum was simply one contribution to that broader set of exchanges.

Denmark: Scenario Workshops

The following example has been adapted from Andersen and Jaeger (2002).

The scenario workshop method was developed in the early 1990s by the Danish Board of Technology (DBT), an independent institution the Danish Parliament established in 1986. The DBT has experimented with and developed participatory methodologies that allow ordinary citizens to be involved in technology assessment. A basic principle of the DBT is that technology assessment should include the wisdom and experience of ordinary citizens or laypeople, integrate the knowledge and tools of experts, respect the political processes and the working conditions of policymakers, and build on the democratic tradition in Denmark.

The DBT's understanding of technology assessment has a background in Danish democratic traditions. As technology becomes more and more integrated into society, influencing more circumstances in life, citizens should have a right to influence its development democratically. This viewpoint initiates a discussion about democracy and technology assessment. As a result, scenario workshops have been used for a variety of issues.

A scenario workshop is designed to find solutions to a problem. It is a local meeting that involves dialogue among four groups of actors: policymakers, business representatives, experts, and ordinary citizens. The participants carry out assessments of technological and nontechnological solutions to the problems, and develop visions for future solutions and proposals for realizing them. A facilitator guides the process. Dialogue among participants with different types of knowledge, views, and experience is central. Various techniques can be employed to accomplish good dialogue and to produce results in the form of identification of barriers, visions, and proposals for action to be taken.

In 1991 the DBT agreed on "sustainable housing and living in the future" as a topic for a new project. The project, it was believed, would benefit from broad consensus on how to develop and transform cities and urban communities to make them ecologically sustainable. The concept of "urban ecology" became a focal point around which the project could formulate more concrete ideas of what was needed in an overall effort toward sustainable development. Urban ecology was defined as the interaction between people and nature in urban areas. To think and act in an ecological way implies saving resources, recycling and reusing products and materials, and returning materials to nature in a nonharmful form. It is concerned with the interaction among different types of technology and various actors, different criteria for assessing technology, different types of knowledge, a wide range of laws and rules from different agencies, various places and levels of action, and several possible solutions. It soon became clear that this project was dealing with an extensive process of societal transition. The project had to address the whole technical infrastructure for managing energy, water, wastewater, and solid waste, as well as the daily life, habits, and values of all the actors involved.

Scenario workshops were conducted in four local communities during 1992. The criteria for choosing the communities were that they have some experience with making a positive effort regarding urban ecology, and that the four communities be of different sizes and levels of urban development. According to the established method, before the workshops took place a set of scenarios was written describing alternative ways of solving the problem. These had to be different with respect to both the technical and organizational solutions described and the social and political values embedded in them. In the workshop the scenarios would be used as visions to provide inspiration for the process.

The workshop process had three principal steps:

1. Commenting on, and criticizing, the scenarios by pointing out barriers to realizing the visions

2. Developing the participants' own visions and proposals

3. Developing local plans of action

The participants first met at "role group" workshops at which participants from the same role group, for example, businesspeople, met in the four localities selected to comment on the scenarios. Reports from these workshops were used as input for the next round of workshops—local workshops, with a mix of members from across each of the four communities. The crosslocal dialogue gave new knowledge on barriers and new ideas on visions to both participants and organizers.

At the local workshops participants were split into "theme groups" according to their experience and interests. The task of each group was to agree on a common vision and produce local action plans for managing energy, water, and waste. The outcome of the whole process was a report and a national plan for urban ecology, which was presented at a public conference in January 1993. Subsequently this was partly implemented by the Danish minister of the environment.

The results of the workshops were threefold:

1. Barriers to urban ecology were identified.

2. Visions were developed.

3. Action plans were proposed.

The results of the workshops in these three areas have played an important role in the Danish debate on sustainable housing and planning during the years since the conference. The following give an idea of some of the changes that have been made because of the workshop:

- In 1993 the minister of environment established a national committee on urban ecology inspired by recommendations from the national action plan.

- In 1995 the Urban Ecology Committee decided to establish a Danish Center of Urban Ecology to support experiments and give advice to those engaged in local activities, and a Green Foundation to finance activities such as those of the Ecological Council and the Association of Green Families.

- The DBT has developed a fund to supply grants for local activities. It has supported hundreds of local meetings with material about urban ecology and money to arrange the meetings.

- The public debate in general has developed scenarios to solve urban ecology problems toward more awareness of the importance of urban ecology principles to be integrated in regulation- and lawmaking.

An evaluation completed by all participants shortly after the project showed that the experience had been an important learning exercise and had paved the way for better dialogue at the local level. However, the DBT has not followed up on the long-term changes resulting from this project in the four communities.

Through the scenario workshop method all the actors contribute the knowledge, vision, and experience they have acquired from local activities to proposals and plans of action on important technology issues. They can all be regarded and defined as experts, because local experience and knowledge are crucial factors in this method. Furthermore, the workshop process tends to bring together people who do not usually engage in dialogue even if they live in the same place.

The scenario workshop method offers a new way of hearing "the voice of the people" and is a supplement to the conventional avenues for participation, such as elections, referenda, and opinion polls. The method cannot claim to express the voice of all the people, but it does offer an opportunity for citizens to present their ideas and opinions in a more open way, which they have the opportunity to structure themselves.

It has been shown that the results from scenario workshops have had some direct effects on decisions taken. More important, though, is their indirect influence, because they give politicians new knowledge about citizens' views of the threats and opportunities of technology, and give citizens new knowledge and awareness. In general, it is difficult to measure and document both the direct and indirect effects of this method.

Local participation may also have a negative aspect, because the results may not be usable at a more general level. More than one workshop process may be needed, as in the case of the Danish urban ecology project, to produce results that can be generalized and used by other local communities or at a national level. This is a question of the availability of resources and time. The success of this method therefore depends on the existence of a body that wants to use the results at the local, national, or even international level.

The scenario workshop method also requires good preparation, planning, and facilitation. If the results are to be used as input for decisionmaking, it may also require that the organizers document the results and present them in a structured way. What becomes increasingly clear from both the Danish experience and initiatives in other countries is that there is one indispensable requirement if success and real change are to take place: the policymakers to whom the results are addressed must be willing to listen and take the results seriously as proposals from the public.

Dams: A Rights-Based Approach

The following discussion is based on World Commission on Dams (2000).

In 1998, through a process of dialogue and negotiation involving representatives of the public, private, and civil society sectors, the World Commission on Dams (WCD) was created. In light of the international debate over large dams, the commission's objectives were to review the development effectiveness of large dams, assess alternatives for water resources and energy development, and develop internationally acceptable criteria, guidelines, and standards, where appropriate, for the planning, design, appraisal, construction, operation, monitoring, and decommissioning of dams.

The commission's 12 members were chosen to reflect regional diversity, expertise, and stakeholder perspectives. The WCD was created as an independent body, with each member serving in an individual capacity and none representing an institution or a country. For two years the commission, together with the WCD Secretariat, the WCD Stakeholders' Forum, and hundreds of individual experts and affected people, conducted a broad and independent review of experience with large dams. This review included public consultations on every aspect of the dams debate and consideration of a large number of submissions. In its report the WCD presented its findings, but also proposed an approach not only to large-dam construction, but to dam development in general. This approach is one based on the recognition of a broad set of human rights and the fact that development often impinges on people's rights, particularly those of the poor. As a result of its review, which was a kind of MSD, the WCD thus developed an improved process framework for governments and donors to adopt for use in the future when considering the creation of a large dam.

As a result of the process of dialogue, study, and reflection, which was an inclusive process that brought all significant players into the debate, the commission

- conducted the first comprehensive global and independent review of the performance of essential aspects of dams and their contribution to development;

- shifted the center of gravity in the dams debate to one focused on investing in options assessment, evaluating opportunities to improve performance and address the legacies of existing dams, and achieving an equitable sharing of benefits in the development of sustainable water resources; and

- demonstrated that the future for the development of water and energy resources lies with participatory decisionmaking using a rights-and-risks approach that will increase the importance of the social and environmental dimensions of dams to a level once reserved for the economic dimension.

The WCD's report found that dams have made a significant contribution to human development, but in too many cases an unacceptable and often unnecessary price has been paid to secure those benefits, especially in social and environmental terms, by people displaced, by communities downstream, by taxpayers, and by the natural environment. Perhaps most significant is that social groups bearing the social and environmental costs and risks of large dams, especially the poor, the vulnerable, and future generations, are often not the same groups that receive the water and electricity services or the social and economic benefits of these. The lack of equity in the distribution of benefits has called into question the value of many dams in meeting water and energy development needs when compared with the alternatives. By bringing to the table all those whose rights are involved and who bear the risks associated with different options for the development of water and energy resources, the WCD created the conditions for a positive resolution of competing interests and conflicts.

The commission's review made it clear that to improve development outcomes in the future a substantially expanded basis for deciding on proposed water and energy development projects is required. All parties should have a complete knowledge and understanding of the benefits, impacts, and risks of large dam projects, and new voices, perspectives, and criteria should be introduced into the decisionmaking process, as well as processes to build consensus. A new paradigm for decisionmaking will improve the outcomes of future decisions. Involving all the stakeholders might bring increased competition for water and thus greater conflict, but it also will lay a foundation for cooperation and innovation.

The work the commission conducted led it to view the controversy within a broader normative framework. This framework builds upon international recognition of human rights, the right to development, and the right to a healthy environment. The WCD decided on five core values that should inform its understanding of the issues:

- Equity

- Efficiency

- Participatory decisionmaking

- Sustainability

- Accountability

The members of the commission believed that these core values are necessary for improved decisionmaking processes that deliver improved outcomes for all stakeholders.

Reconciling competing needs and entitlements is the single most important factor in understanding and resolving the conflicts associated with large-scale development projects. The approach developed by the commission—recognizing rights and assessing risks (particularly when rights are at risk)—offers a means to apply the WCD's core values to decisionmaking. Clarifying the rights context of a proposed project is an essential step in identifying the various claims and entitlements that the project or its alternatives might affect. It is also a necessary step in determining the stakeholder groups entitled to participate in the decisionmaking. The assessment of risk adds an important dimension to understanding how, and to what extent, a project may affect people's rights. This requires seeing risk as something faced not only by governments and developers, but by those affected by a project and by the environment as a public good. Once all the parties whose rights are at stake have been brought to the table, a transparent process and negotiated outcome are possible.

Based on its core values and rights-based perspective, the WCD developed seven strategic priorities for the process of decisionmaking on dams. These priorities were designed to provide guidance in translating the rights-based approach into practice and to help development processes move from a traditional, top-down, technology-focused approach to ones that are inclusive of those the project will affect and of normative considerations.

1. Gaining public acceptance through recognizing rights, addressing risks, and safeguarding the entitlements of all groups of people affected. Decisionmaking processes and mechanisms are used that enable informed participation by all groups of people, and result in the demonstrable acceptance of key decisions.

2. Assessing options in a comprehensive and participatory fashion through all stages of a project based on the needs of all groups. The option selected is based on an assessment of the full range of policy, institutional, and technical options.

3. Improving existing dams and addressing the outstanding social and environmental issues. Management must adapt to changing circumstances continuously over the project's life.

4. Sustaining rivers and livelihoods for ecosystems and human communities dependent on them. Options assessment and decisionmaking around river development prioritize the avoidance of impacts, followed by the minimization and mitigation of harm to the health and integrity of the river system.

5. Recognizing the entitlements of affected peoples through joint negotiations to produce mutually agreed-upon and legally enforceable mitigation and development provisions, and sharing benefits.

6. Ensuring compliance, public trust, and confidence by requiring governments, developers, regulators, and operators to meet all commitments, regulations, criteria, guidelines, and project-specific negotiated agreements made for the planning, implementation, and operation of dams.

7. Sharing rivers among and within countries for peace, development, and security through collaborative and innovative means.

These priorities were not intended as a blueprint. Instead the commission recommends that they be used as the starting point for discussions, debates, internal reviews, and reassessments of existing procedures and for an assessment of how these procedures might need to change. The experience of the commission in a dialogue among parties from different backgrounds illustrates that common ground can be found without stakeholders' compromising their interests and values. But it also shows that all the parties concerned must commit to the process if the issues are to be resolved.

Summary and Conclusions

Multistakeholder processes aim to address multidimensionality and complexity, the intrinsically politically charged issues of allocation of rights to resources, and the distribution of benefits and costs associated with technological change. This chapter has argued that success in reconciling deeply held positions and arriving at consensus on future paths hinges on the extent to which four basic factors are addressed:

1. The degree to which relevant parties are involved in discussions and negotiations

2. The extent to which accurate scientific information is brought forward

3. The quality and depth of linkages with official decisionmaking bodies

4. The degree to which fairness and efficiency are embraced as evaluation criteria

The three examples provided in the previous section illustrate the inherent context-specificity of multistakeholder processes. A unified, fully portable approach (model) does not exist, suggesting the need for contingent approaches that are cognizant of institutional and political details, and of the opportunities and constraints these details may imply. The examples also illustrate the decisiveness of the interactive effects of the nature of available evidence, the social and political context within which policy change is debated and implemented, and the facilitative mechanisms at hand.

A key recognition relates to the thin and incomplete nature of information about and understanding of the institutional and political context within which science and technology policy is made in developing countries, especially in Africa, and especially with respect to biotechnology policy. Biotechnology is a tool to be used to meet societal goals. Investments in alternative policy approaches are therefore best viewed in relation to particular constraints on achieving such goals. Again, the degree of understanding of how such investments might address key constraints is thin and incomplete. The need for contingent approaches is therefore especially great for multistakeholder processes dealing with biotechnology in Africa.

Appendix: Alternatives for Process Design

Following is a select list of methods that could be used for "deliberative and inclusionary processes." This list has been adapted from Holmes and Scoones 2000.

Area/Neighborhood Forums

Such forums are concerned with the needs of a particular geographically defined area or neighborhood. Meeting regularly, they may deal with a specific service area (e.g., planning or housing) or with a full range of local services and concerns. Area forums may or may not have dedicated officers attached to them. They may have a close link with relevant ward council members or with council members responsible for the service areas under discussion. Membership may be set or open. If there is a formally established membership (e.g., consisting of representatives of tenants or community associations in the area), members of the public may be allowed to participate in an open discussion at meetings.

Citizens' Juries

A citizens' jury is a group of citizens (chosen to fairly represent the local population) brought together to consider a particular issue set by a local authority. Citizens' juries receive evidence from expert witnesses, and cross-questioning can occur. The process may last up to four days, at the end of which a report is drawn up setting out the views of the jury, including any differences in opinion. Juries' views are intended to inform council members' decisionmaking.

Citizens' Panels

Research panels. Research panels are bodies made up of a large sample of a local population (500–3,000 participants) that are used as a sounding board by an organization in the public sector. They are part of a form of opinion research that tracks changes in opinions and attitudes over time. Members are recruited either through the mail or by telephone. Such panels have a standing membership, a proportion of whom will be replaced regularly and who will be consulted at intervals. Participants are asked regularly about different issues over a period of time. An example is the People's Panel on public services for the U.K. central government.

Interactive panels. Interactive panels also have a standing membership that may be replaced over time, but they consist of small groups of people who meet regularly to deliberate on issues. An example would be a health panel.

Community Issues Groups

The community issues group takes the focus group (described later) as its starting point, then attempts to introduce the core elements of deliberation. A group of up to 12 people come together up to five times to discuss a designated issue in depth. Each meeting lasts for up to two and a half hours. The first meeting has a similar format to that of a focus group; participants discuss an issue from their current knowledge base. In subsequent meetings information is introduced so that their knowledge of the subject area is gradually increased. By the final meeting participants have become more informed and the opinions they express have moved beyond their automatic initial responses toward more thoughtful and anchored judgments (for example, the public vision of U.K. health service).

Consensus Conferences
Consensus conferences involve a panel of laypeople who develop their understanding of technical or scientific issues in dialogue with experts. A panel of between 10 and 20 volunteers are recruited through advertisements. A steering committee is set up with members chosen by the sponsors. The panel members attend two weekend meetings at which they are briefed on the chosen subject and identify the questions they want to ask at the conference. The conference lasts for three or four days and gives the panel a chance to ask experts any outstanding questions. The conference is open to the public, and the audience can also ask questions. Then the panel members retire and, independent of the steering committee, prepare a report that sets out their views on the subject. Copies of the report are made available to the members of the conference audience, and panel members present key sections to the audience.

Consensus Participation
The framework used in consensus participation involves six activities. First, stakeholder analysis involves identification of the relevant stakeholder groups. Second, stakeholder targeting involves bringing all stakeholders to a position in which they are able to negotiate with other stakeholders on a more equitable basis. Third, external stakeholder assessment involves investigating the policies, legislation, and activities of the government and other institutional stakeholders that may constrain or promote local actions. Fourth, community participatory assessments enable local people to identify their resource uses, assess perceived conflicts and concerns, and plan community strategies. Fifth, participatory preparatory workshops bring all the stakeholders together to cover a series of specific crosscutting issues. Participants produce a series of position statements that provide the basis for subsequent discussions. Sixth is the policy planning forum, where facilitators manage negotiations between stakeholders to build consensus and reach agreement on policies and projects. Seventh, participatory monitoring and evaluation take place using criteria agreed upon during the policy planning forum.

Deliberative Opinion Polls

These polls measure informed opinion on an issue. A deliberative poll examines what members of the public think when they have had the time and information necessary to consider the matter more closely. These polls usually involve 250–600 participants. A baseline survey of opinions and demographics is carried out, and the participants in the poll are then recruited to resemble the wider group in terms of both demographics and attitudes. Often briefing begins before the event by means of written information. Participants' views on a given subject are measured before the poll begins and again once it has finished. Changes in opinion are measured and incorporated into a report. Deliberative polls are often conducted in conjunction with television companies.

Electronic Democracy

Two forms of electronic democracy are informal on-line discussions and formal consultations using on-line debates. Informal discussions enable participants to share knowledge through informal writing aimed at a real audience, and they leave a record of conversations that can be referred to later. Because all communications must be in writing, contributions are often thoughtful, with everybody on an equal footing. Discussions are similar to face-to-face conversations but are a sequence of messages or postings that are "asynchronous" because contributors typically do not participate at the same time. Formal debates are moderated and focus on specific questions to be argued for or against. Moderators provide content relevant to the debate and facilitate discussion. In an online environment, formal debate can take place by dividing participants into teams and assigning each team a specific argument. Debates may take the form of heightened discussions in which participants discover and investigate concepts and conflicts within a topic or issue. Some participants may be assigned the task of respectfully disagreeing with others' stated points of view.

Focus Groups

Focus groups are similar to citizens' juries in that they bring citizens together to discuss a specific issue. Focus groups do not need to be representative of the general population, and may involve a particular group of citizens only. Discussions may focus on the specific needs of that group, on the quality of a particular service, or on ideas for broader policy or strategy. Focus groups do not generally call expert witnesses, and meetings typically last between one and two hours only, usually involving only 12 people.

Future Search Conferences

A future search conference is a two-day meeting at which participants attempt to create a shared community vision of the future. It brings together those with the power to make decisions and those affected by decisions to try to seek agreement on a plan of action. The process is managed by a steering group of local people representing key sections of the community. About 64 people are recruited who are asked to form about eight stakeholder groups within the conference. They take part in a structured two-day process in which they move from reviewing the past to creating ideal scenarios for the future. Each of the stakeholder groups explains its vision, and then a shared vision is explored. The conference ends with the development of action plans. Self-selected action groups develop projects and commit themselves to action toward their vision.

Innovative Development

Innovative development is a methodology consisting of four participatory steps. First, an "action map" is formulated. This is a systematic vision for action toward an attainable and desired future that reflects the consensus of participants. Second, an estimate is made of the distance from the current situation to the attainable future and of the capabilities that are available. Third, a study is made of "potentialities," which includes the systematic identification and evaluation of each of the prospective actions. Fourth, a plan for action is designed. All methodological steps are carried out through the participation of "relevant actors or stakeholders" who are convoked by an appropriate and legitimate authority.

Issue Forums

Issue forums are ongoing bodies that hold regular meetings focusing on a particular issue (e.g., community safety or health promotion). They may have a set membership or may operate on an open basis, and they are often able to make recommendations to relevant council committees or to share in decisionmaking processes.

Multicriteria Mapping

Multicriteria mapping (MCM) is a method that attempts to combine the transparency of numerical approaches with the unconstrained framing of discursive deliberations. The technique involves deciding on a subject area, defining the basic policy options, selecting the participants, conducting individual interviews (two- to three-hour sessions in which additional options are selected, evaluative criteria are defined, options are scored, and relative weighting is given to criteria), quantitative and qualitative analysis is conducted by researchers, feedback on preliminary results is provided for participants, deliberation between participants takes place, and, after the final analysis, a report is produced.

Participatory Rural Appraisal or Participatory Research and Action

Participatory rural appraisal or participatory research and action (PRA) is a family of approaches, methods, and behaviors that enable poor people to express and analyze the realities of their lives and conditions and themselves in order to plan, monitor, and evaluate their actions. In PRA outsiders act as catalysts for local people, enabling them to decide what to do with the information and analysis that they generate. PRA methods are similar to those used for rapid rural appraisal (see following).

Planning for Real

Planning for real is a hands-on planning process first developed in the 1970s as an alternative to traditional planning meetings. Using models and cards, it can be used to address many issues such as traffic, community safety, conditions of housing stock, and environmental improvements. Planning for real exercises are often initiated by a neighborhood or residents' group. Material is provided by the Neighborhood Initiatives Foundation to help people embark on a neighborhood survey to identify problems and issues. A three-dimensional model of a neighborhood is prepared by all sections of the community. The model is moved around the area to places accessible to the community. A planning for real event is an open meeting that focuses attention on the model. At the meeting "Movable options" cards are used to identify problem areas and discuss how they might be solved. The event is followed by workshops to prioritize options and identify responsibility for each action.

Rapid Rural Appraisal

Rapid rural appraisal (RRA) consists of data collection by outsiders (researchers or practitioners who are not members of the community or group with whom they interact) through the use of methods that include participant observation, semi-structured interviews, and visual techniques (e.g., maps, matrices, trend lines, and diagrams).

Service User Forums

Service user forums are ongoing bodies that meet on a regular basis to discuss issues relating to the management and development of a particular service (e.g., an older people's day care center, a leisure center, or park). Such forums may have a set membership or operate on an "open" basis. They may have the power to make recommendations to specific council committees or even to share in decisionmaking processes.

Stakeholder Decision Analysis

Stakeholder decision analysis is a method of combining a deliberative procedure (e.g., discussion and negotiation between stakeholders) with systematic multicriteria decision analysis. Deliberations among stakeholders elicits criteria that reflect underlying value judgments. The criteria are weighted according to their relative importance during a series of workshops. Each social or environmental issue of concern is then scored against its criterion. Weighted scores are summed to give a final score. This process can focus discussions between stakeholders, facilitating networking and partnership building, promoting negotiation, and avoiding confrontation. Because it is open and transparent, it is seen to be fair. The outcomes gain legitimacy from the procedure followed.

Visioning Exercises

A range of methods (including focus groups) may be used within a visioning exercise, the purpose of which is to establish participants' "vision" of the future and the kind of future they would like to create. Visioning may be used to inform broad strategy for a locality, or it may have a more specific focus (as in environmental consultations for Local Agenda 21).

For a description of other methods that could be used for participatory environmental policymaking, see Holmes and Scoones 2000.

References

Andersen, I.-E., and B. Jaeger. 2002. Danish participatory models scenario workshops and consensus conferences: Towards more democratic decision-making. In *Pantaneto Forum*, April 6. http://www.pantaneto.co.uk/issue6/andersenjaeger.htm.

Carlson, C. 1999. Convening. In *The consensus building handbook: A comprehensive guide to reaching agreement*, ed. L. E. Susskind, S. McKearnan, and J. Thomas-Larmer. Thousand Oaks, CA: Sage Publications.

Consensus Building Institute. 2001. *How to conduct a conflict assessment.* Boston.

———. 2002. *Key facts in the joint fact-finding (JFF) process.* Boston.

Holmes, T., and I. Scoones. 2000. *Participatory environmental policy processes: Experiences from north and south.* Institute of Development Studies (IDS) Working Paper 113. Sussex, England: IDS. Full report available at http://www.ids.ac.uk/ids/bookshop/wp/wp113.pdf.

IIED (International Institute for Environment and Development). 2002. *Breaking new ground.* http://www.iied.org/mmsd/finalreport/ or through Earthscan Publishers, London, England.

IIED/WBCSD (International Institute for Environment and Development / World Business Council for Sustainable Development). 2001. *What is MMSD producing?* Brochure. http://www.iied.org/mmsd/mmsd_pdfs/what_is_mmsd_producing.pdf.

Montana Consensus Council. 1998. *Phases of building agreement.* Helena.

Pimbert, M. P., and T. Wakeford. 2002. Prajateerpu: A citizen's jury/scenario workshop on food and farming futures for Andhra Pradesh, India. London: International Institute for Environment and Development.

Scoones, I., and J. Thompson. 2003. Participatory processes for policy change: Reflections on the Prajateerpu e-forum. http://www.iied.org/docs/sarl/eforumsumm.pdf.

Susskind, L. E. 1999. A short guide to consensus building. In *The consensus building handbook: A comprehensive guide to reaching agreement,* ed. L. E. Susskind, S. McKearnan, and J. Thomas-Larmer. Thousand Oaks, CA: Sage Publications.

UNDESA (United Nations Department of Economic and Social Affairs). 1997. Agenda 21. http://www.un.org/esa/sustdev/documents/agenda21/english/agenda21toc.htm.

World Commission on Dams. 2000. Executive summary. http://www.earthscan.co.uk/dams/summary.htm.

Chapter 3

Agricultural Biotechnology, Politics, Ethics, and Policy

Julian Kinderlerer and Mike Adcock

The aim of this chapter is to address the policy, regulatory, and ethical issues surrounding agricultural biotechnology. The chapter provides background on the shaping of policy and on regulatory frameworks within the European Union and the United States, among others, as well as outlining the global framework in which all countries have to operate. In addition it summarizes the United Nations–led initiative to assist developing countries to implement biosafety frameworks devised by a specific country for that country. The chapter also highlights the ongoing debate in the areas of environmental protection, public perception and acceptance, and intellectual property rights.

The most important reason for addressing the policies of the European Union and the United States on genetically modified organisms rather than the policies of other nations is that they are very different in concept—although in practice, once the regulatory system has been triggered their formal treatment of such organisms is very much the same.

The introduction of a new technology such as agricultural biotechnology may depend on the perceived balance between the benefits of the technology and the potential risks to the environment and human health. This chapter aims to put forward the arguments and issues related to the potential benefits of agricultural biotechnology against a background of perceived risks, but it does not seek to provide the answers.

Policy

Different uses of modern biotechnology to produce transgenic organisms elicit varying reactions in most countries. The use of genetic modification to provide medicines is not as controversial as the genetic modification of crops for human consumption. Often the genetic modification of animals (especially reproductive cloning) is considered less acceptable than the modification of plants. Modification of the germ line in humans, for example, is often considered immoral or contrary to *ordre public*.[1] This is made explicit in Article 6 of Directive 98/44/EC of the European Union (European Union 1998a).

Many opinion polls indicate that the public discriminates markedly between uses of biotechnology. Using such technology in medicine and horticulture/floriculture is often found to be acceptable, whereas the genetic modification of crops for food use and the modification of animals and humans are less acceptable. Hallman et al. (2002, p. 26) report: "While most Americans say they would be in favour of at least some genetically modified food products, and nearly two-thirds believe that genetically modified foods will benefit many people, more than half (56 percent) say that the issue of genetic modification causes them great concern."

History

It may be useful to provide some historical background on the many issues that arise in response to the use of modern biotechnology. Policy on the safe use of biotechnology sets precedents. It is often the case that safety legislation is introduced because an accident has occurred and systems need to be put into place to ensure that such an accident does not recur. The possible risks of modern biotechnology were recognized at the very beginning of its use, and steps were taken to ensure that it was used safely.

The potential uses of genetic modification[2] were obvious from the moment that researchers first identified the techniques that enabled the transfer of genes from one organism to another unrelated organism. A committee (the Ashby Committee) established by the government of the United Kingdom reported in 1975 that genetic manipulation techniques would provide "substantial though unpredictable benefits" and that "application of the techniques might enable agricultural scientists to extend the climatic range of crops and to equip plants to secure their nitrogen supply from the air" (United Kingdom, Working Party, 1975). A meeting of scientists using the new recombinant DNA technology at Asilomar, California, in February 1975 produced a set of guidelines for the use of biotechnology. The formal goals of the meeting included identifying the "possible risks involved for the investigator and/or others" and "the measures that can be employed to test for and minimize the biohazards so that the work can go on" (Wright 1994, p. 145). In the view of the Ashby Committee, the benefits of the new technology far outweighed the risks if suitable precautions were put in place (United Kingdom, Working Party, 1975; emphasis added).

Although in many countries the public has been fearful of the introduction of the products of this technology, governing bodies have not been as reticent, and have recognized both the benefits that may arise from its use and the risks that it theoretically poses. On May 13, 1993, the Parliamentary Assembly of the Council of Europe passed Recommendation 1213 on developments in biotechnology, for which it indicated there were many wonderful prospects, but for which there were also many concerns (Council of Europe, Parliamentary Assembly 1993).[3] The Council of Europe includes many countries of central and eastern Europe as well as those of the affluent European Union.[4] The resolution noted that the gene pool has been widened far beyond the limits of sexual compatibility to encompass the possibility of transferring genes from almost any organism to others. Among the many uses of biotechnology it identified were increasing agricultural outputs (or reducing inputs), replacing chemical herbicides and insecticides or more efficient targeting of these products, increasing the use of plants in industry, reducing the response of crop plants to stress, and even cloning meat animals "for particular markets or to form embryo banks to maintain genetic diversity." The resolution noted that significant drawbacks might result from the application of the new biotechnology. The possibility of new diseases was raised, as were the potential environmental effects of transgenic organism.[5] Many of the benefits have been effected, although many people do not realize that many vaccines, pharmaceuticals, and food additives (such as chymosin and ascorbic acid) are the products of modern biotechnology.[6]

The Cartagena Protocol on Biosafety (Secretariat of the Convention on Biological Diversity 2000) was agreed to by the members of the Conference of the Parties to the Convention on Biological Diversity (CBD) in 2000 in Montreal.[7] This came after years of negotiation and argument, with the misgivings of many parties, but in an atmosphere that had changed from that which had prevailed at the time the negotiations had started in 1995 at the second meeting of the parties to the CBD in Jakarta. Article 19(3) of the Convention on Biological Diversity (Secretariat of the Convention on Biological Diversity 1992) had required parties to consider the possibility of adding to the convention a protocol that addressed the use (and primarily transboundary movement) of living modified organisms that might have an adverse impact on biological diversity.[8] Eight years later Europeans were no longer accepting modern biotechnology; products had disappeared from the shops, and there was a gloom and distrust in many countries not observed elsewhere. Few if any products derived using modern biotechnology are now available in Europe (Royal Society of the United Kingdom 2002, para. 2). In North America, farmers adopted transgenic organisms with little opposition, and products derived from them have been in shops for more than five years.

The developing countries wanted far more to be included in the protocol than they were able to get, including many more safeguards. The producer countries fought hard to ensure that, insofar as it was possible, few if any controls would be applied, particularly to commodity goods. The size of the commodity market alone, they argued, made it difficult to contemplate a regime that required what amounted to "visas" at country entry points.

The Cartagena Protocol required 50 ratifications to come into force. In accordance with its Article 36, the protocol was opened for signature at the UN office in Nairobi during the fifth ordinary meeting of the Conference of the Parties to the Convention on Biological Diversity in Nairobi, Kenya, May 15–26, 2000. It remained open for signature at the UN headquarters in New York from June 5, 2000, to June 4, 2001. By that date the protocol had received 103 signatures. The Cartagena Protocol entered into force on September 11, 2003, some 90 days after receipt of the 50th instrument of ratification. African countries that have ratified the protocol are Algeria, Botswana, Burkina Faso, Cameroon, Djibouti, Egypt, Ethiopia, Gambia, Ghana, Kenya, Lesotho, Liberia, Madagascar, Mali, Mauritius, Mozambique, Namibia, Niger, Nigeria, Rwanda, Senegal, Seychelles, South Africa, Togo, Tunisia, Uganda, the United Republic of Tanzania, and Zambia. Zimbabwe signed the protocol in 2001 but has not yet ratified it. Most of these countries do not yet have the legal systems in place to implement the requirements of the protocol.

The need for specific legislation in regard to the use of genetically modified organisms was never presumed even though it was recognized that regulation was needed from the earliest days of the use of this technology. The United Kingdom had regulated the genetic "manipulation" of microorganisms starting in 1978, and by 1983 it had a full set of legally binding regulations in place. The United States, on the other hand, had specified guidelines (the National Institutes of Health [NIH] guidelines) that identified the manner in which such organisms should be used by those funded by the NIH.

In 1986 the U.S. government published its Coordinated Framework for the Regulation of Biotechnology (U.S. Office of Science and Technology Policy 1986), which described the "comprehensive federal regulatory policy for ensuring the safety of biotechnology research and products." The document set forth some of the assumptions on which it was based, as follows: "Existing statutes provide a basic network of agency jurisdiction over both research and products; this network forms the basis of this coordinated framework and helps assure reasonable safeguards for the public. This framework is expected to evolve in accord with the experiences of the industry and the agencies." The laws that already existed in the United States regulated the uses of specific products, such as foods or pesticides. It had been thought that genetically modified organisms posed no new risks that could not be covered using the existing system. But according to the document, "This approach [that offered by the framework] provides the opportunity for similar products to be treated similarly by particular regulatory agencies" (pp. 23302–23350).

The framework describes the rationale for its development:

> The underlying policy question was whether the regulatory framework that pertained to products developed by traditional genetic manipulation techniques was adequate for products obtained with the new techniques. A similar question arose regarding the sufficiency of the review process for research conducted for agricultural and environmental applications. . . . Upon examination of the existing laws available for the regulation of products developed by traditional genetic manipulation techniques, the working group concluded that, for the most part, these laws as currently implemented would address regulatory needs adequately. For certain microbial products, however, additional regulatory requirements, available under existing statutory authority, needed to be established." (U.S. Office of Science and Technology Policy 1986, p. 23302)

The U.S. government decided to identify the various tasks needed to regulate biotechnologies and clearly indicate the agency and even the law that would be used to ensure that these technologies were used safely. Other countries did not (at the time) have a range of environmental, food, drug, and safety legislation in place that permitted effective use of existing legislation. In the United States it was decided that jurisdiction over the many different biotechnology products would be determined by their use rather than by the manner of their production, just as was the case for traditional products (see Table 3.1).

Regulatory Systems

Guidelines or regulations were quickly introduced in some countries, particularly to protect those who might come into contact with the modified organisms. In the United Kingdom the first regulations were introduced in 1978; in the United States the NIH guidelines were implemented soon after the 1975 meeting at Asilomar and applied to work funded through grants received from the NIH. Initially the "regulations" applied primarily to work in laboratories, because that was the

Table 3.1 Agencies responsible for approval of commercial biotechnology products under the U.S. Coordinated Framework for the Regulation of Biotechnology

Products	Agencies
Foods and food additives	Food and Drug Administration (FDA)
Human drugs, medical devices, and biologics	FDA
Animal drugs	FDA
Animal biologics	Animal and Plant Health Inspection Service (APHIS)
Other contained uses	Environmental Protection Agency (EPA)
Plants and animals	APHIS, Food Safety and Inspection Service (FSIS), FDA
Pesticide microorganisms released into the environment	EPA, APHIS
Other microorganisms, intergeneric combinations	EPA, APHIS
Intrageneric combinations: pathogenic source organisms	
1. Agricultural use	APHIS
2. Nonagricultural use	EPA, APHIS
Intrageneric combinations: no pathogenic source organisms	EPA
Nonengineered pathogens	
1. Agricultural use	APHIS
2. Nonagricultural use	EPA, APHIS
Nonengineered pathogens	EPA

Source: U.S. Office of Science and Technology Policy 1986, p. 23304.

only place in which the work could progress. They were aimed at the protection of those individuals who had access to the laboratories and attempted to ensure that the work was contained and that workers were protected from the hazards posed by the modified organisms. It was only in the late 1980s that the introduction of modified organisms into the environment became really feasible. At first it was expected that these releases would mainly be of microorganisms, but as methods capable of modifying plants became available and efficient it was clear that most environmental releases would be of plants. Very few modified microorganisms have been released. Many countries have decided to implement different systems of regulation for organisms intended for use in containment and those released into the environment. Organisms are considered to be used in containment when they are used in industrial plants and in processes for manufacturing in which the organisms themselves are not intended to be marketed or exposed to the "open" environment.

Most countries in the southern African region are considering the frameworks necessary for a regulatory system to ensure the safe use of modern biotechnology or have already enacted legislation. South Africa initially regulated transgenic organisms[9] through a voluntary system, but since 1997 has had legislation in place to ensure that in South Africa modified organisms are used safely (South Africa 1997). So far it is the only country in the region that has permitted the commercial use of any transgenic plants. According to an article in the *Financial Gazette*,

"Zimbabwe was the second country after South Africa to come up with biosafety regulations; was the first to come up with an institutional framework and is one of the few countries to have graduate training in biotechnology" (Nyathi 2002). Namibia was part of a pilot project funded by the United Nations Environment Program (UNEP) and the Global Environment Facility (GEF) that permitted 18 countries to start the process of regulating biotechnology, and it is now one of 12 countries financed by the GEF to implement the biosafety frameworks that have been devised for the country. Kenya, Uganda, and Zambia were also among the countries that participated in the pilot project, and Kenya and Uganda are among the 12 now implementing their frameworks with significant funding from the GEF. Botswana, Lesotho, Mozambique, Rwanda, Zimbabwe, and other countries in the region are currently being funded through a project implemented as a follow-up to the pilot project, which assists countries to design frameworks to ensure the safety of biotechnology.[10]

Countries have chosen to use a variety of triggers for the regulation of biotechnology. In Europe it is using modern biotechnology as defined in the directives (European Union 1998b and 2001)[11] that triggers the regulatory process. In the United States, because previously existing law is used the trigger tends to be the use of organisms that are pests—plant pests, for example—in the manufacture of the new organism if the Department of Agriculture is to be involved. Canada has chosen to use a concept of novelty to trigger the regulatory process. Many analyses have suggested that once the process is started, the risk assessment and management processes are very similar in the various countries.

Environmental Policy in Relation to Genetically Modified Organisms

All the countries that are participating in GEF-funded projects have signed the Cartagena Protocol (Secretariat of the Convention on Biological Diversity 2000), which specifically requires regulation in relation to the transboundary transfer of living modified organisms that may have adverse effects on the conservation and sustainable use of biological diversity, also taking into account risks to human health.[12] Those participating in the "implementation" projects have also ratified or acceded to the protocol or have agreed to do so. They are also all party to the Convention on Biological Diversity (Secretariat of the Convention on Biological Diversity 1992), whose Article 8(g) requires that they institute national frameworks in order to "establish or maintain means to regulate, manage or control the risks associated with the use and release of living modified organisms resulting from biotechnology which are likely to have adverse environmental impacts that could affect the conservation and sustainable use of biological diversity, taking also into account the risks to human health." The provisions of the Cartagena Protocol extend only to those organisms resulting from modern biotechnology that might cause potential adverse effects to the conservation and sustainable use of biodiversity. Human health has "then" to be taken into account. However, when designing a regulatory system for biosafety, it is legitimate to ensure safety of the environment and human health in general, with the needs for the protocol forming a subset within the regulatory system. It seems likely that any attempt to link the protection of human health to legislation that primarily addresses biodiversity would not be acceptable to most legislatures.

Countries have understood that in this instance biosafety means primarily protection of the environment, and that the release of living modified organisms needs be regulated in order to protect the environment.[13] Safety concerns are not, however, limited to the impact of these organisms on the environment, and regulatory systems that attempt to ensure human and animal health are often different from those set in place for environmental protection. The European Novel Food Regulation agreed to in 1997 (European Union 1997)[14] provided extensive risk assessment and management for the use of genetically modified organisms or products derived from them in foods. This has now been replaced by Regulation 1829/2003 (European Union 2003a), which applies to food or feed produced using genetic modification. It provides "the basis for ensuring a high level of protection of human life and health, animal health and welfare, environment and consumer interests in relation to genetically modified food and feed, whilst ensuring the effective functioning of the internal market." It sets out the EU procedures for authorization and supervision of genetically modified products and contains provisions for the labeling of genetically modified food and feed (Article 1). Regulation 1830/2003 (European Union 2003b) addresses the traceability and labeling of genetically modified organisms and the traceability of food and feed products that have been derived from such organisms. These two regulations significantly extend the requirements that were put in place under the previous regulation. In particular, products derived from genetically modified organisms but in which the modification is not detectable (neither the DNA nor any protein produced due to the action of the inserted gene is present) must be labeled to indicate their derivation.

Precaution

Scientific data can be collected at many sites around the world that can provide an insight into the manner in which a product of biotechnology may interact with its environment when released into a particular environment. When data are not available or when a country believes that its environment is different from that in which the organism was tested, field testing may be required before the organism is released or placed on the market. Where data are "knowable," further experimentation will provide information that may address concerns as to the likely behavior of the organism in a particular environment. However, because of the inherent variability of biological systems, such information may fall into the "not knowable" category; that is, no amount of information collected may be able to provide more than increased precision in determining the variability of the organism's behavior. Further experimentation will not provide any assurance that the organism will (or will not) affect the environment in an unacceptable manner. This "precautionary principle" or approach is invoked in order to address the absence of data. It is usually taken to refer to Principle 15 of Agenda 21 (UNCED 1992), agreed to in Rio de Janeiro in 1992: "In order to protect the environment, the precautionary approach shall be widely applied by States according to their capabilities. Where there are threats of serious or irreversible damage, lack of full scientific certainty shall not be used as a reason for postponing cost-effective measures to prevent environmental degradation."

Many cases of serious environmental degradation have made governments change their perception of environmental protection. These cases have also affected the public's perception of the environment. Outbreaks of disease in animals and humans due to perceived lack of care or to environmental pollution have had a significant effect on an appreciation of both known and potential risks to the environment and to human health and on public acceptance that these potential problems need to be addressed. According to the Organization for Economic Cooperation and Development (OECD 2002, p. 7): "The use of precaution cannot be limited to approving an action or process, or prohibiting it, but implies managing various levels of risk and uncertainty, and taking the appropriate measures at each level." A risk may vary significantly depending on the level of activity or the likelihood that an organism may persist and establish itself in the environment. The organism's interrelationship with other actions or processes or with other organisms with which genetic material may be exchanged may also require caution in analyzing the potential risk.

Annex III of the Cartagena Protocol (Secretariat of the Convention on Biological Diversity, 2000) identifies the principles for scientific risk assessment that member countries need to address when considering living modified organisms that might have adverse effects on biological diversity, also taking into account the impact on human health. It provides, inter alia, that "lack of scientific knowledge or scientific consensus should not necessarily be interpreted as indicating a particular level of risk, an absence of risk, or an acceptable risk."

This precautionary principle (or approach) has attracted many and various interpretations; for many it means that if the science is unknown and there is a risk of environmental damage, one should not proceed. Caution dictates that it implies that when there is doubt over the safety of an action, that action should not be taken until evidence is available that the steps to be taken will not have disastrous consequences for the environment. The concern in relation to transgenic organisms is due to the possibility that once an organism is in the environment it will be virtually impossible to recall and, because of its property of replication, it will not decay over time; indeed its numbers may increase disastrously. Others interpret this as an injunction to proceed with caution, considering each release into the environment on a case-by-case basis and probably also proceeding step-by-step, with small field trials preceding larger ones and the results analyzed before proceeding to commercial unfettered release (if ever). According to the Commission of the European Communities (2000, p. 1), recourse to the precautionary approach "presupposes that potentially dangerous effects deriving from a phenomenon, product or process have been identified, and that scientific evaluation does not allow the risk to be determined with sufficient certainty." A Canadian discussion document reflects the following view: "Decision making about risks in the context of a precautionary approach is further complicated by the inherent dynamics of science. Even though scientific information may be inconclusive, decisions will still have to be made to meet society's expectations that risks be addressed and living standards maintained" (Government of Canada 2001a). Scientists may be concerned that the 'principle' is used to stifle research, innovation, and competition. The Commission of the European Communities further states:

Where action is deemed necessary, measures based on the precautionary principle should be, *inter alia:*

- *proportional* to the chosen level of protection,
- *nondiscriminatory* in their application,
- *consistent* with similar measures already taken,
- *based on an examination of the potential benefits and costs* of action or lack of action (including, where appropriate and feasible, an economic cost/benefit analysis),
- *subject to review,* in the light of new scientific data, and
- *capable of assigning responsibility for producing the scientific evidence* necessary for a more comprehensive risk assessment. (2000, p. 4, para. 6; emphasis in original)

The World Trade Organization (WTO) Agreement on the Application of Sanitary and Phytosanitary (SPS) Measures (WTO 1994b) reflects precaution in Article 5.7, which allows members to adopt SPS measures where relevant scientific evidence is insufficient. If members are to use precaution, they should meet four specific conditions:[15]

- The measure must be provisional, although no time limit is set.

- It must be adopted on the basis of "available pertinent information."

- An attempt must be made "to obtain the additional information necessary for a more objective assessment of risk."

- The measure must be reviewed within a reasonable period of time.

The use of precaution requires that a number of major considerations be taken into account. The Canadian discussion document provides a starting point for defining policy in relation to precaution:

1. "The decision-making process for managing risks always requires sound and rigorous judgment" where "[J]udgment means determining what is a *sufficiently* sound or credible scientific basis, what *follow-up* activities may be warranted, and *who* should produce a credible scientific basis."

2. "To reduce significant scientific uncertainty and improve decision making, the precautionary approach usually includes follow-up activities such as research and scientific monitoring." However, it has to be noted that in many instances the collection of data may increase the precision of determination of variation, rather than provide data which permits the reduction of uncertainty. Monitoring can only provide assurance that expected events occur, and events predicted not to occur are not observed. Unexpected, unpredictable, indirect and delayed effects on the environment are by their nature difficult if not impossible to monitor." (Government of Canada 2001b, p. 4)

The arguments around the precaution principle are serious, for they have directly affected the policy decisions of many countries. In Europe the use of precaution in relation to transgenic organisms is taken to require case-by-case and step-by-step approaches to risk. This way of interpreting precaution is built into the Cartagena Protocol, which also requires a case-by-case process in assessing risk (Secretariat of the Convention on Biological Diversity 2000, Annex III.6).

Public Opinion

The controversy over the use of modern biotechnology has centered primarily on commercial release into the environment rather than on use in laboratories for research, contained use in industry, use in the production of pharmaceuticals and veterinary products, or even use in field trials. Protesters have, however, chosen to attack and destroy fields in which organisms are being tested. The industrial use of genetically modified organisms that may be the major use of modern biotechnology now and in the future. The Eurobarometer surveys show that considerable discrimination among the public (at least in Europe) in relation to the various uses of modern biotechnology (Eurobarometer 2000): "Europeans continue to distinguish between different types of applications, particularly medical in contrast to agrifood applications" (Gaskell, Allum, and Stares et al. 2003). Support for genetically modified crops and foods declined and opposition increased over the period between 1996 and 1999; from 1999 to 2002 there was almost no change in levels of support or opposition. European attitudes toward six applications of biotechnology (Gaskell, Allum, and Stares et al. 2003) indicate the discrimination that has been observed. The results displayed in Figure 3.1 indicate how discriminating the public is. For example, genetically modified food is considered risky, morally unacceptable, and not to be encouraged, yet genetically modified crops (much to the surprise of the researchers) are considered useful but risky, but their use is seen as morally acceptable and a slight majority favors their use! In a survey of Canadians it was found that "a total of 47.7% of Canadians consider the presence of GMOs [genetically modified organisms] in foods to be dangerous for human health while 20.7% feel they are not dangerous" (31.6 percent did not express an opinion) (Leger Marketing 2001).

The European public debate resulted in rejection of modern biotechnology, which in 1998 had the effect of influencing the main distribution companies to remove these products from European shelves. In the United States, there appeared to be little rejection, which the U.S. government attributed to the openness of the American system: "In 1994 approximately 7,000 acres were planted under 593 USDA [U.S. Department of Agriculture] field-test authorizations, compared to 57,000 acres under 1,117 authorizations in 2001. The first biotechnology-derived crops were commercialized in 1996 and, in 2001, approximately 88 million acres were planted in the United States and 130 million acres were planted world-wide" (U.S. Office of Science and Technology Policy 2002, pp. 50578–50580). Argentina,

Figure 3.1 European attitudes toward six applications of biotechnology, 2002

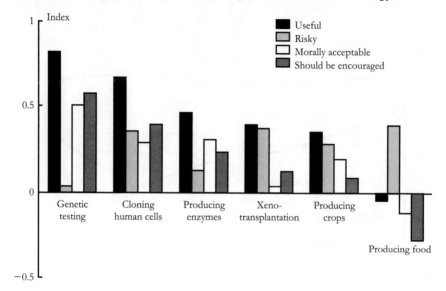

Source: From Gaskell, Allum, and Stares et al. 2003, p. 13.
Note: The response alternatives for these questions were on a four-point scale (definitely agree, tend to agree, tend to disagree, and definitely disagree) and were recoded by the authors as −1.5 to +1.5 (on the *y*-axis here) in order to show the midpoint of zero in the figure.

Canada, and Mexico are the only other countries that have made significant use of modern agricultural biotechnology, although many other countries, including Australia and South Africa, are starting to increase their use of living modified organisms in agriculture. China has approved a small number of transgenic varieties of cotton and expects to proceed to the commercial production of modified rice in the next two years. The latest Eurobarometer survey of European attitudes toward technology (Gaskell, Allum, and Stares et al. 2003) indicated that Europeans had recovered their faith in technology, including biotechnology, but the results, shown in Figure 3.2, may simply indicate that the de facto moratorium on the commercialization of plants manufactured using genetic modification techniques has taken the subject out of the public consciousness.

In the United States, according to Hallman and associates, the "American public's position on the acceptability of genetic modification of food is decidedly . . . undecided." Some 58 percent of Americans either strongly approve or somewhat approve of creating hybrid plants using genetic modification, while 37 percent disapprove (Hallman et al. 2002, p. 20).

Figure 3.2 European optimism about technologies, 1991–2002

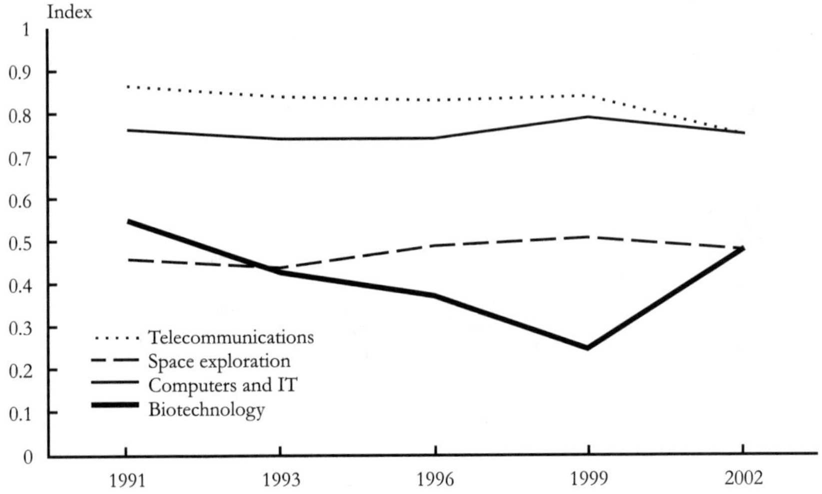

[Line chart with Index on y-axis (0 to 1) and years 1991, 1993, 1996, 1999, 2002 on x-axis. Legend: Telecommunications (dotted), Space exploration (dashed), Computers and IT (solid thin), Biotechnology (solid thick).]

Source: Gaskell, Allum, and Stares et al. 2003.

Many developing countries are fearful of the impact of agricultural biotechnology. Zambia and Zimbabwe, for example, have been wary of permitting food aid that includes transgenic maize to come into the country, even though many of their people are starving. This reluctance relates to concerns about the safety of the food when it forms a very high percentage of intake and also relates to the possible disappearance of major markets if crops are "contaminated" with transgenic material. Zimbabwe has accepted transgenic maize when it has been milled.

What is happening in Europe is significant, because it has a direct bearing on what can be done in developing countries. In the first instance, the concerns being expressed by Greenpeace, Friends of the Earth, Christian Aid, and even the British Medical Association[16] create a groundswell against the use of this new technology. Can it be right to introduce these "untested" technologies in developing countries when public "informed" opinion is so virulently opposed to their use in Europe? When even statutory bodies like the nature conservation organizations in Britain and France reject modern biotechnology because of its predicted negative effect on the environment, are developing countries to embrace them? The United Nations Environment Program's International Guidelines (UNEP 1995) and the Cartagena Protocol (Secretariat of the Convention on Biological Diversity 2000) require that the public be informed and educated about biosafety, but the virulent reaction against this technology in Europe directly affects its public image more easily than does a reasoned argument for the safe use of the technology. In Britain, during the first nine months of 1999 there were a continual series of press reports "implying that eating GM food would lead to all sorts of serious diseases" (United Kingdom, House of Commons 1999, para. 29).

The attention paid by the media to foods produced using modern biotechnology has been sustained over a long period and has been almost totally hostile. The coverage has stressed the technology rather than the products. The rejection of genetically modified foods by many European supermarkets and food producers has had an impact on the production and growing of genetically modified crops that have to be exported to one of the largest food markets in the world.[17] The possibility of growing rice modified so that it produces vitamin A is a wonderful prospect for nutrition in the many countries that depend on rice as a primary food. However, the produce cannot be exported as well, producers will be reluctant to grow it! Concern over the impact of genetically modified crops on the environment has been the primary concern, but fears about the long-term safety of eating modified foods and about the speed of entering the unknown have sent powerful messages to the public (Burton 1999). An article in a Christian Aid paper asks, "Are GM crops the next in a long line of inappropriate products to be dumped on poor countries?" It continues: "GM crops are irrelevant to ending hunger; the new technology puts too much power over food into too few hands; and too little is done to help small farmers grow food in sustainable and organic ways. . . . It is tempting to see biotechnology in agriculture as a clean neutral science, simply transferring progress from the laboratory to the field, improving the lot of everyone. This is illusory. All technologies are embedded in specific economic and social systems and have different costs and benefits" (Burton 1999).

This response to the new technology in Western Europe cannot easily be dismissed through assertions by scientists that there is negligible risk or that permits to market transgenic foods and crops (in particular) should be based solely on risk assessments that are science-based. If all the scientific information were available and a consensus among scientists could be achieved that the impact of such foods on the environment is minimal, it would be possible to argue for a totally science-based risk assessment process. An Irish consultation paper (Republic of Ireland, Department of the Environment and Local Government, 1999) expresses some of the problems: the concerns about potential environmental and human health effects arise due to an absence of familiarity with the regulatory systems; the technology is complex and developing rapidly; "there is little experience on the interaction of GMOs with their surrounding environment"; the information being provided to the public is probably inadequate, particularly in relation to labeling to allow choice; the use of antibiotic resistance marker genes is thought to be inimical to their use in human and veterinary medicine; and the use of herbicide-tolerant crops might increase the use and build-up of herbicides in the environment.

In 2000 the Council of Europe Parliamentary Assembly once again looked at the use of modern biotechnology (and, in particular, the patenting of genes and gene fragments) and resolved: "Public opinion should be more strongly involved in political decision-making as regards scientific and technological choices and scientists should be encouraged to engage more in public debate" (Council of Europe, Parliamentary Assembly 2000).

Policy on involving the public has evolved in many different ways. Article 23 of the Cartagena Protocol (Secretariat of the Convention on Biological Diversity 2000) requires that countries engage their publics in decisionmaking both at the policy level and when considering individual applications for use of modern biotechnology:

1. The Parties shall:
   (a) Promote and facilitate public awareness, education and participation concerning the safe transfer, handling and use of living modified organisms in relation to the conservation and sustainable use of biological diversity, taking also into account risks to human health. In doing so, the Parties shall cooperate, as appropriate, with other States and international bodies;
   (b) Endeavour to ensure that public awareness and education encompass access to information on living modified organisms identified in accordance with this Protocol that may be imported.
2. The Parties shall, in accordance with their respective laws and regulations, consult the public in the decision-making process regarding living modified organisms and shall make the results of such decisions available to the public, while respecting confidential information in accordance with Article 21.

Even for countries with a history of involving their publics in the decisionmaking process this is not easy; for those not used to direct public involvement it may be much more difficult.

Science-Based Decisions

Many have argued that decisions on the use of living modified organisms must be based on science; policy may be defined when designing the system that is applied to individual applications, but the applications should be considered only in the light of this policy.

Decisions are usually made by governments based on advice received from a number of sources. The risk assessment procedure, at the very least, should be science-based. This is made very clear in the Cartagena Protocol (Secretariat of the Convention on Biological Diversity 2000). Article 15 states: "Risk assessments undertaken pursuant to this Protocol shall be carried out in a scientifically sound manner." A report by the Royal Society of Canada (2002, para. 3) asserts that "scientific assessments must inform policy decisions but cannot pre-empt them, and that public opinion must be taken into account throughout." The report writers continue: "We believe that the public debate about GM food must take account of wider issues than the science alone. We also wish to stress the importance of informing debate with sound science." Article 23 of the Cartagena Protocol (Secretariat of the Convention on Biological Diversity 2000) requires public involvement in the decisionmaking process, and Article 26 allows for specific socioeconomic issues to be taken into account in the process: "The Parties, in reaching a decision on import under this Protocol or under its domestic measures implementing the Protocol, may take into account, consistent with their international obligations, socioeconomic considerations arising from the impact of living modified organisms on the conservation and sustainable use of biological diversity, especially with regard to the value of biological diversity to indigenous and local communities."

Unlike Canada, the European Union, and the United States, the vast majority of developing countries may not have expertise directly employed by the government in the vast array of disciplines needed to perform a complete risk assessment of transgenic organisms. The data needed to assess likely environmental degradation or impact may not be available in many countries. In such countries a different approach may be needed, whereby an applicant requesting a permit for the use of a transgenic organism must perform a detailed risk assessment—possibly even performing field tests in an appropriate environment—and submit the resulting data for audit to the government, rather than the government performing the risk assessment. Most scientists may feel more confident in auditing a detailed assessment than attempting the assessment themselves. Applicants could also be expected to design their own risk management, consultation, and monitoring procedures, with input from government-appointed assessors when appropriate. There is an obvious danger inherent in this approach, however, for the government's lack of trust in those applying to release organisms to provide all the necessary information may mitigate against the acceptance of the risk assessment. Can applicants be trusted to provide all the necessary information? If a decision is made to use an audit rather than a direct risk assessment by the government, it is important that the scientists involved in the audit be able to ask for further information and be able to identify gaps in the approach taken by the applicant.

Risk assessment of genetically modified organisms is largely based on the concept of familiarity, or of "substantial equivalence," which assumes that all the characteristics of the modified organism are those of the host organism except for the specific characteristics introduced. It is actually difficult to identify other ways of approaching the problem of identifying risk. But the Royal Society of Canada (2001) and the Royal Society of the United Kingdom (2001) have both indicated dissatisfaction with "substantial equivalence." Can the approach be justified when stress tolerance, modification of metabolism, or production of pharmacologically active compounds really begins?

Crop varieties developed through conventional plant-breeding techniques not involving modern biotechnologies are not generally tested for their safety. Rather they have to meet plant variety registration requirements that identify whether they are distinct from those currently on the market, uniform, and stable. These traditional methods use (primarily) crossing selection and back-crossing processes to select a desired characteristic and remove inadvertently introduced extra characteristics that initially accompany the introduced trait. These mechanisms introduce new and numerous gene combinations. If toxins or allergens are known to occur in these crops (e.g., glucosinolates in canola, glycol-alkaloid accumulation in potatoes), the new variety is normally tested to ensure that the level of toxin or allergen is no greater than the range that is normally observed for that substance. Interactions of traits introduced by traditional methods with other characteristics of the plant are normally ignored until they can be proven to make the variety unusable. According to the Royal Society of Canada (2001; emphasis in original): "The implicit assumption behind this methodology is that, even where a breeding-derived novel trait is involved, *new combinations of existing genes operating within highly selected germplasm are not expected to generate harmful outcomes.*"

The concept of substantial equivalence was introduced for use with transgenic crops. It was first described in a report of the OECD (1993) that suggested: "If a new food or food component is found to be substantially equivalent to an existing food or food component, it can be treated in the same manner with respect to safety." The World Health Organization published a report (WHO 1995) in which the concept of substantial equivalence as a decision threshold was promoted as the basis for safety assessment decisions concerning GMOs (Royal Society of Canada 2001, p. 179).

The Royal Society of the United Kingdom (2002) has said that substantial equivalence can be considered in three ways:

- The GM foodstuff might be regarded as substantially equivalent to its conventional counterpart both toxicologically and nutritionally. . . .

When a product has been shown to be substantially equivalent, no further safety assessment is required.
- It might be substantially equivalent apart from certain defined differences. Sometimes the GM food product includes the components deliberately introduced by genetic modification. In this case the GM food product might be regarded as "substantially equivalent to its conventional counterpart except for a small number of clearly defined differences." Assessment is then limited to examining the implications of the difference(s), perhaps by testing the novel components of the GM plant in isolation.
- The GM product might be regarded as not substantially equivalent to its conventional counterpart, or there might not be a suitable reference available for comparison. The product will then need a highly detailed safety assessment taking all the properties of the modified foodstuff and determining by direct measurement where necessary the impact on human health and the environment.

Many countries are deciding that using the term *substantial equivalence* is misleading. It suggests that if substantial equivalence is demonstrated, no further assessment need be done. A report by the Food and Agriculture Organization (FAO) and the World Health Organization (WHO) (2000) says that there was a "mistaken perception that the determination of substantial equivalence was the end point of a safety assessment rather than the starting point." In 2002 the Royal Society of the United Kingdom recommended: "Safety assessments should continue to consider potential effects of the transformation process. The phenotypic characteristics to be compared between foods derived from GM plants and their conventional counterparts should be defined. It may not be necessary or feasible to subject all GM foods to the full range of evaluations but those conditions that have to be satisfied should be defined" (Royal Society of the United Kingdom 2002, p. 10).

Intellectual Property Rights and Ethics

Many arguments have been made for and against the use of intellectual property rights in relation to modern biotechnology. According to a resolution of the Parliamentary Assembly of the Council of Europe: "The patent system, as a system for the protection of intellectual property, is an integral part of the market economy and therefore can be a driving force for innovation in many technological questions" (Council of Europe, Parliamentary Assembly, 1999). The same resolution notes that "living organisms are able to reproduce themselves even if they are patented, and in view of this special quality of living organisms the scope of a patent is difficult to define, which makes it nearly impossible to find a balance between private and public interests." The resolution also notes that there are ethical concerns related to the use of patents on living systems:

9. The Assembly considers that monopolies granted by patent authorities may undermine the value of regional and worldwide genetic resources and of traditional knowledge in those countries that provide access to these resources.
10. It considers that the aim of sharing the benefits from the utilisation of genetic resources within this broader view does not necessarily require patent-holding but requires a balanced system for protecting both intellectual property and the "common heritage of mankind."
11. It also considers that the many outstanding questions regarding the patentability and the scope of protection of patents on living organisms in the agro-food sector must be solved swiftly taking into account all interests concerned, not least those of farmers and developing countries. (Council of Europe, Parliamentary Assembly, 1999)

Over the last few decades the global trading importance of biotechnology has been recognized. As a result, concerted and concentrated efforts have been made to protect the results of research and development involving genetic material. The result of this has been the extension of intellectual property protection to most forms of biological material. The trade importance of biological information has been underlined by the adoption of the Agreement on the Trade Related Aspects of Intellectual Property Rights (TRIPS) within the World Trade Organization (WTO 1994c). This agreement requires states party to the agreement to provide protection for all types of inventions irrespective of the field of technology. The aim of the agreement is to ensure that all member states provide effective and appropriate intellectual property protection and protect intellectual property rights by the appropriate enforcement mechanisms. The agreement sets down the minimum standards of protection.[18] Article 27(2) of the TRIPS agreement permits countries to exclude from patentability those inventions whose commercial exploitation may be contrary to *ordre public* or morality. Countries may exclude from patentability "diagnostic, therapeutic and surgical methods for the treatment of humans or animals." More important, Article 27(2) allows members to exclude from patentability innovations produced in order to protect animal or plant life or health or to avoid serious damage to the environment, and Article 27(3) provides for exclusion from patentability of "plants and animals other than micro-organisms, and essentially biological processes for the production of plants or animals other than nonbiological and microbiological processes. However, Members shall provide for the protection of plant varieties either by patents or by an effective sui generis system or by any combination thereof" (WTO 1994c).

What constitutes sui generis protection for new plant varieties is not defined; hence countries are free to adopt a system that ensures intellectual property protection for plants. One option is for countries to implement UPOV (the International Union for the Protection of New Varieties of Plants), which was established by the International Convention for the Protection of New Varieties of Plants,[19] but Simon Walker believes that "this form of protection has been criticized for focusing too much on the rights of plant breeders, and too little on the rights of those using the seeds—farmers" (Walker 2001).

Although member states are obliged to provide protection systems, those "inventing" new products do not need to obtain that protection. The rights apply only in the country in which the inventors have chosen to invoke protection. In most African countries many of the biotechnology inventions have not been protected through patent rights and can legally be used as if in the public domain. It is only when products developed using patent protected materials or methods are exported into countries where protection is offered that the rights of the inventor must be respected.

There is an underlying assumption that the introduction of an intellectual property system will result in a dramatic increase in the innovative capacity of the private sector while allowing the public sector to become more self-financing. This may be true to an extent in countries with a substantial research capacity, but it is unlikely to be the case in developing countries, where the research and development sector is not as strong. A "Northern" intellectual property system may provide an incentive, but there may be limited local capacity to exploit it. Even when technologies are developed, firms in developing countries can seldom bear the costs of acquisition and maintenance of rights, much less those of enforcement (especially in those countries where substantial earnings may be realizable). The costs of establishing an infrastructure to support an intellectual property rights regime may be substantial, and mechanisms for the enforcement of such rights are costly both to government and to private stakeholders.

If a country has made a policy commitment to implement a rights system, perhaps the best way to proceed would be to look at the systems in Europe and the United States and adapt them to local and cultural needs. The required patent system would need to balance the costs and benefits against local needs and requirements. Those responsible for the implementation of such a system should examine whether there might be a need to

- raise the standard of the granting criteria of novelty, inventiveness, and industrial application to ensure that the reward of the patent is consummate with the benefit to society;

- widen the range of subject matter that can be excluded from patentability;

- provide an effective compulsory licensing system;

- include an exclusion of patentability on the grounds of "morality" similar to that found in Article 53(a) of the European Patent Convention; and

- consider the suitability of other forms of protection to encourage local innovation, such as utility models.

There is real concern about the use of intellectual property law in developing countries, particularly in relation to health care, but also in relation to what is emotively called biopiracy or bioprospecting. In May 2000 the revocation by the European Patent Office of a patent on a neem[20] product was undoubtedly a victory for India and developing countries. However, individual legal action is no substitute for a legally enforceable integrated approach to bioprospecting.

Pharmaceutical companies worldwide are interested in finding new and alternative therapies and have widened their search to include traditional medicines and practices largely based on medicinal plants endemic to developing countries. Many traditionally used herbal medicines may have real therapeutic properties. If a company takes traditional knowledge as the starting point for a search for new pharmaceuticals and extracts the active product, it is entitled to a patent on the extracted product even though it cannot replace the traditional product itself. Developing countries are thus faced with the acute dilemma of having their valuable indigenous wealth taken away and exploited commercially by the resource- and technology-rich transnational pharmaceutical companies.

Bioprospecting is not found just in the area of pharmaceuticals. In northwest Mexico, yellow beans have been cultivated for centuries as they are the staple diet of many Mexicans. In 1994 John Proctor, the owner of a small-seed company, POD-NERS, LLC, bought a bag of commercial bean seeds in Mexico and took them back to the United States. Proctor planted the yellow beans in Colorado and allowed them to self-pollinate. When yellow beans were selected over several generations, a segregating population resulted in which the color of the beans is uniform, stable, and changes little by season. In 1996 Proctor applied for a U.S. patent that was granted in 1999.[21] With the patent granted, Proctor has an exclusive monopoly on yellow beans and can exclude the importation, sale, offer for sale, make, use for any purpose, including drying edible or propagation of any yellow bean exhibiting the yellow shade of the Enola beans.

Customs officials at the U.S.-Mexico border are now inspecting beans, searching for any patent-infringing beans being imported into the United States. Because of this bean alone and the threat of patent infringement, some export sales of yellow Mexican beans have dropped over 90 percent. This has also had an affect on the market for other nonyellow beans, as often the beans are not separated and yellow patent-infringing beans are mixed with nonyellow beans. As agriculture is the primary source of employment and livelihood for the people of northwest Mexico, this patent has had a serious effect on farmers in that area. Although farmers can still grow and sell the beans in Mexico, they can no longer export them to markets in the United States without paying royalties to the patent holder.

The International Center for Tropical Agriculture is legally challenging the patent, arguing that the patent claims are invalid because they fail to meet the requirements related to novelty and nonobviousness and disregard available prior art. The opposition proceedings have been slowed by the filing of new claims by POD-NERS, and no decision has been made as yet.

One extremely important lesson can be learned from what many people feel is an example of bioprospecting at its worst. In the United States,[22] according to 35 USC 102(a) an invention cannot be "known or used in this country, or patented or described in a *print publication* in this or a foreign country"(emphasis added). Therefore, mere use in Mexico without printed publication is insufficient to show a lack of novelty. Hence the need to document genetic resources, as we will discuss later.

Membership in the WTO requires that countries have in place an effective intellectual property regime. However, the simple implementation of the TRIPS agreement in national law is insufficient to protect a country's genetic resources, as Article 27(3b) is inadequate to meet their protection requirements. What is required is the enactment of legislation that incorporates the framework of current agreements and negotiations—TRIPS, along with the requirements of the Convention on Biological Diversity and the International Treaty for the Protection of Plant Genetic Resources.

In November 2001, at the WTO ministerial conference in Doha concerns of this sort resulted in a statement and an agreement to find a solution to some of these pressing problems before the end of 2002. No agreement has yet been reached. The Doha statement (WTO 2001a) recognized

- the gravity of the public health problems afflicting many developing and least developed countries, especially those resulting from HIV/AIDS, tuberculosis, malaria and other epidemics. . . .
- the need for the WTO Agreement on Trade-Related Aspects of Intellectual Property Rights (TRIPS Agreement) to be part of the wider national and international action to address these problems. . . .
- that the TRIPS Agreement does not and should not prevent Members from taking measures to protect public health.

The final item continues: "Accordingly, while reiterating our commitment to the TRIPS Agreement, we affirm that the Agreement can and should be interpreted and implemented in a manner supportive of WTO Members' right to protect public health and, in particular, to promote access to medicines for all."

The ministers also recognized that compulsory licensing to produce drugs was not an option for many of the developing countries, and that other solutions would have to be found for many of these countries. Hence developing countries should consider the manner in which they implement the various agreements in order to protect their people and their resources, paying heed to the following:

1. Developing countries should enact appropriate biodiversity protection legislation including benefit sharing consistent with Article 8j[23] of the Convention on Biological Diversity (Secretariat of the Convention on Biological Diversity 2000) and access to genetic resources (covered in Article 15).

2. The TRIPS agreement requires not that countries institute a patent regime for plant material, but that they create a sui generis system for protection of the plant intellectual regime (Walker 2001). The replacement system could be designed to protect extant varieties that are in the public domain as well as new plant varieties and to provide for the needs of the country taking into account, for example, the communitarian approach to property that is often part of the culture of developing countries as well as the needs for innovation.

3. Developing countries may need to document and catalogue their biological assets not only to ensure protection but also to ensure future collaboration and exploitation. States have sovereign rights over their biodiversity and are responsible for conserving their biological diversity and for using their biological resources in a sustainable manner.[24] Article 3 of the Convention on Biological Diversity (Secretariat of the Convention on Biological Diversity 2000) reads: "States have, in accordance with the Charter of the United Nations and the

principles of international law, the sovereign right to exploit their own resources pursuant to their own environmental policies, and the responsibility to ensure that activities within their jurisdiction or control do not cause damage to the environment of other States or of areas beyond the limits of national jurisdiction."

Many are concerned with the way in which intellectual property (IP) protection has been used in many countries. As Walker writes:

> The balance in many IP systems seems to be shifting too far in favour of technology producers. Negotiations over IPRs have been powerfully influenced by industry lobby groups and are being driven by concerns of trade liberalization and international investment between developed countries. The legitimate technological and developmental objectives of developing countries—generally technology users—are not being given due consideration. This shift in the ownership and control of information, and the resulting boon to private investors, has been called an "information land grab." (Walker 2001)[25]

Ethical Issues Raised by Modern Biotechnology

In May 1999 the Nuffield Council on Bioethics, an independent organization in the United Kingdom, published a major report titled "Genetically Modified Crops: The Ethical and Social Issues" (Nuffield Council on Bioethics 1999). The executive summary of the report states: "The application of genetic modification to crops has the potential to bring about significant benefits, such as improved nutrition, enhanced pest resistance, increased yields and new products such as vaccines. The moral imperative for making GM crops readily and economically available to developing countries who want them is compelling."

Many have argued that transgenic crops will assist in the task of providing enough food in the right places and at the right times to retain, as far as possible, the way of life of those who desperately need food. However, in order to do so, it is essential that the crops that are modified and the genes inserted be chosen with the needs of those who are hungry in mind. To suggest that the modified crops currently available are primarily anything other than products designed for industrialized farming is clearly wrong; however, the technology has been used where it was possible in the early stages of its development. The development of new uses that really do benefit those who are needy is imperative if this technology is to benefit the poor. In the words of the Council of Europe's Parliamentary Assembly (2000):

"It is increasingly important to include ethical considerations centred on humankind, society and the environment in deliberations regarding developments in biotechnologies, life sciences and technologies and their applications."

Natural and Unnatural Products

Many perceive the use of genetically modified organisms in the environment as equivalent to "playing God," as an unnatural act that should not be done. There is a deep-rooted belief in many societies that tinkering with nature, or the industrialization of nature, is unacceptable. This argument will be at least as strong in African societies as it is in Europe. Many hold the view that tampering with nature is inherently wrong, that we have "dominion" over nature,[26] which implies a responsibility to look after and protect nature rather than own it.

The idea that genetic modification "that could not happen naturally" is wrong is held by many people even though it is not often clearly enunciated. Many argue that this concept precludes any selective approach that results in improved crop plants, for by using such approaches we are playing God. Others argue that it is only that which could not have happened without human intervention that is unacceptable. Even if modification itself is seen as acceptable, there might be religious objections that would mean that the resulting organism would be unacceptable. For example, insertion into foods of genes derived from a pig could arguably be unacceptable to those whose religion precludes the use of products derived from this "unclean" animal.

Any discussion based on objections to playing God is generally not accessible to logical argument. Respect for such beliefs usually involves ensuring that there are mechanisms in place to permit believers to choose not to use such products. According to the Nuffield Council (1999, para. 6.7): "Proponents of the technology citing practical benefits may have an intrinsic value system that views science and progress as good things in themselves, and opponents may be analysing risks from a world-view that questions the rightness of technological progress."

The Principle of Justice

One of the most important issues that we need to recognize is that many different groups within a society have competing rights and fears. We need to attempt to balance these needs. "For example," writes the Nuffield Council (1999, para. 1.20), "if protecting the rights of consumers by providing adequate labeling was very expensive and was generally agreed to do nothing to prevent harm, most people would say that upholding the right to know would not be worth the loss of value to producers, particularly if the producers were poor. Conversely, if informative but inexpensive labeling was desired by the majority of consumers, it would probably command wide public support." The principles at stake are not complex, but their implementation is. Securing a consensus is complicated by the fact that producers have an interest in exaggerating the difficulty of complying with new regulations, and pressure groups have an opposite interest in exaggerating the public demand for them. Questions about where the balance of burden and benefit is to be struck are the subject of everyday political debate.

This principle of justice poses many questions that need to be addressed. Is this new technology likely to increase the gap between the rich and the poor, both within countries (particularly in the developing countries) and between developed and developing countries? Are the products produced by the technology able to provide for those who really need them, the poor? Will the technology generate wealth for the society as a whole that can assist those who need it? If the technology is more efficient and will provide more food but at the expense of some who farm traditionally, is it acceptable?[27] According to the Nuffield Council (1999, para. 1.23), "GM crops are currently vulnerable to questions about their real usefulness and to questions about who benefits."

Economic and Social Benefits and Risks: The Principle of General Welfare

Of necessity biotechnology has to be applied for the benefit of human beings, society, and the environment. These beneficiaries are not necessarily the same, for the benefit to human beings may be at the short- or long-term expense of the environment. There is a presumption that the "acceptability" of the risk must include an improved quality of life, perhaps as we develop better (or more) food, better health, and an environment that is improved in a sustainable manner. Human usage of the environment in the 10,000 years of our exploitation of nature has been relatively benign. In the last 100 years, however, we have made rapid and possibly irretrievable changes to the environment, including the excessive use of fossil fuels relative to their replacement, excessive use of water, production of greenhouse gases, and even a huge increase in the human population. Humans are no longer in harmony with their environment, and we have to be aware of the effect on the environment. Whereas a primary goal of technology was once the pursuit of happiness (and the greatest good), we now have to pursue sustainability.

These concerns are human-centered. Many of those who live in southern Africa are suffering from severe malnutrition, and drought is wreaking havoc with and on the environment. If the application of modern biological techniques can result in food products that can better survive drought and heat, and can also provide more food in the right places at the right times, there are clear benefits that can result from its use. It is axiomatic that food is essential for our survival. According to the FAO (2001, p. 3), "Both formal ethical systems and ethical practices in every society presume the necessity of providing those who are able-bodied with the means to obtain food and enabling those who are unable to feed themselves to receive food directly." And, in the words of the *Rome Declaration on World Food Security* (FAO 1996b):

> We consider it intolerable that more than 800 million people throughout the world, and particularly in developing countries, do not have enough food to meet their basic nutritional needs. This situation is unacceptable. Food supplies have increased substantially, but constraints on access to food and continuing inadequacy of household and national incomes to purchase food, instability of supply and demand, as well as natural and man-made disasters, prevent basic food needs from being fulfilled. The problems of hunger and food insecurity have global dimensions and are likely to persist, and even increase dramatically in some regions, unless urgent, determined and concerted action is taken, given the anticipated increase in the world's population and the stress on natural resources.

It is clear that we need to promote access to the genetic resources for food and agriculture for farmers, farming communities, and consumers.

Human health is important in this context. Health is improved when hunger is eliminated and the quality of food is improved. Healthy people are empowered in that they are able to participate in society and are better able to live meaningful lives. The FAO constitution identifies the need to raise levels of nutrition, secure improvements in the efficiency of production and distribution of all food and agricultural products, and better the conditions of those who live in rural areas.

For most consumers in developed countries the choice of whether to eat genetically modified foods is not an ethical issue. To eat genetically modified food would not be wicked, even if the individual was concerned as to its safety. However, if that food was proscribed by the society as (for example) not being *halal,* or kosher, not giving the people the ability to identify the food as proscribed would be unethical. When people are starving and a technology can help to provide them with more and nutritionally better food, but it is not made available, an ethical issue is at stake.

The industrialization of agriculture is an issue in many African countries, for it takes away the traditional structures of society and substitutes a more individualist system that may cause harm. This industrialization might arguably help in providing more and better food at the cost of disrupting traditional belief systems and modifying the way of life of many in rural areas, which may result in a situation in which less food will be available where and when necessary.

The agreement setting up the WTO (WTO 1994a) tried to balance the many conflicting issues that this principle requires:

> Relations in the field of trade and economic endeavour should be conducted with a view to raising standards of living, ensuring full employment and a large and steadily growing volume of real income and effective demand, and expanding the production of and trade in goods and services, while allowing for the optimal use of the world's resources in accordance with the objective of sustainable development, seeking both to protect and preserve the environment and to enhance the means for doing so in a manner consistent with their respective needs and concerns at different levels of economic development.

The WTO and its disputes resolution system has placed the freedom to trade above environmental concerns, but there is recognition of the importance of environmental concerns.

The WTO (2001b, p. 47) outlined some of the issues it would have to address: "If one country believes another country's trade damages the environment, what can it do under the terms of the WTO agreements? Can it restrict the other country's trade? If it can, under what circumstances? At the moment, there are no definitive legal interpretations, largely because the questions have not yet been tested in a legal dispute either inside or outside the WTO." When both countries are party to an international environmental agreement, their dispute may be able to be addressed through that agreement. If one of the countries is not a party to the agreement, it is not yet possible to decide what the implications might be. It will depend on the obligations placed on the member country by the treaty and by the specifications identified in the agreement in regard to relations between parties and nonparties. If neither country involved in the dispute is party to an environmental agreement (or if there is no agreement relating to that issue), WTO rules apply. They have been interpreted to mean that trade restrictions cannot be imposed on a product purely because of the way it has been produced and that any one country cannot impose its standards on another.

Sustainable Development

In 1987 the Brundtland Report of the World Commission for the Environment and Development, also known as *Our Common Future,* considered the need to ensure that economic development was achieved without the depletion of natural resources. The report asserted that it is necessary to provide for the future without harming the environment. Published by an international group of politicians, civil servants, and experts on the environment and development, the report provided a key statement on sustainable development:

> It is in the hands of humanity to make development sustainable, that is to say, seek to meet the needs and aspirations of the present without compromising the ability of future generations to meet their own. The concept of sustainable development implies limits—not absolute limits, but limitations that the present state of technology or social organisation and the capacity of the biosphere to absorb the effects of human activities impose on the resources of the environment—but both technology and social organisation can be organised and improved so that they will open the way to a new era of economic growth. The Commission believes that poverty is no longer inevitable. Poverty is not only a malaise in itself. Sustainable development demands that the basic needs of all are satisfied and that the opportunity of fulfilling their expectations of a better life is extended to all. A world where poverty is endemic will always be susceptible to suffering an ecological or any other kind of catastrophe. (Bruntland 1987)

According to the online *Encyclopaedia of the Atmospheric Environment* (Buchdahl and Hare 2000), "The report highlighted three fundamental components to sustainable development: environmental protection, economic growth and social equity. The environment should be conserved and our resource base enhanced, by gradually changing the ways in which we develop and use technologies. Developing nations must be allowed to meet their basic needs of employment, food, energy, water and sanitation. If this is to be done in a sustainable manner, then there is a definite need for a sustainable level of population. Economic growth should be revived and developing nations should be allowed a growth of equal quality to the developed nations."

This is an important policy statement; it provides for an approach to our environment that must inform the manner in which crops are produced and land is used.

Autonomy, Dignity, Integrity, and Vulnerability

Human autonomy and dignity need to be respected. Article 2 of the United Nations Educational, Scientific, and Cultural Organization (UNESCO) Universal Declaration on the Human Genome and Human Rights (1997)[28] states:

(a) Everyone has a right to respect for their dignity and for their rights regardless of their genetic characteristics.
(b) That dignity makes it imperative not to reduce individuals to their genetic characteristics and to respect their uniqueness and diversity.

Article 6 reads: "No one shall be subjected to discrimination based on genetic characteristics that is intended to infringe or has the effect of infringing human rights, fundamental freedoms and human dignity." Governments are expected to treat the deeply held convictions of their citizens with respect: they have to pursue policies that can command a general consensus even where some views cannot be accepted because they are in direct contradiction with others (Nuffield Council on Bioethics 1999, sect. 1.09). Animals and the natural world are also entitled to respect for their integrity and vulnerability (Nielsen and Faber 2002, p. 12).

There are also concerns that the new technology will lead to exploitation of those living in the "developing" countries. For instance,

- monopoly control of chemicals used in agriculture and of seeds that allow plants to resist these chemicals might be exploitative and place a strain on the economies of developing countries, and

- major changes in social structures might sequentially affect the types of agriculture and needs for distribution of foods and food products.

Just Distribution of Benefits and Burdens

Ethical use of biotechnology requires just distribution. This is particularly important in the context of developing countries, for it has been argued that for obvious reasons most of the products derived from modern biotechnology are being introduced by private companies that have an obligation to maximize earnings for their shareholders, and that therefore the products are aimed at markets that can best pay for their use. If the technology simply increases the divide between rich and poor, can it be ethical? This question will have to be addressed through public and private funds that attempt to provide for those who cannot purchase the new products.

The most important means of providing aid to those living in countries that rely on subsistence agriculture is to ensure the provision of adequate food and clean water. Important benefits may accrue from the provision of technological expertise. It has been argued that the manner in which agricultural resources are distributed should be equitable. Many conflicting arguments have been offered about the equitable distribution of food and farmland between the rich and poor, both in developed and developing countries. According to Gary Comstock (2000), the need to redistribute land to the people of Zimbabwe and to dispossess those who had taken the land during the colonial past was seen as part of an equitable redistribution within Zimbabwe. Comstock also addresses the role of the industrialization of agriculture:

> Most of the world's poor are small tenant farmers. In order to increase the standard of living of these farmers, the governments of many developing countries adopted in the 1970s the policy of "industrializing" agriculture; making their farmers over in the image of large successful farmers in more developed countries. During the green revolution of the 1960s and 70s, countries such as India, Costa Rica, and Nigeria increased the efficiency of farmers' yields by borrowing money from international lending agencies such as the World Bank. The funds were used to extend credit to farmers who in turn were taught to buy high yielding varieties of seeds (such as rice, wheat, and maize) and to use the necessary accompanying technologies: mechanical implements (tractors) and synthetic chemicals (herbicides and pesticides). Many farmers flourished and nations that once were importing grain became self-sufficient in certain crops.

A majority of the world's resource-poor farmers are women. Worldwide, women produce more than 50 percent of all the food that is grown. In many developing countries, this percentage is much higher. For instance, it is estimated that women produce 80 percent of the food grown in sub-Saharan Africa, 50 to 60 percent of that in Asia, 46 percent of that in the Caribbean, 31 percent of the food grown in north Africa and the Middle East, and about 30 percent of that in Latin America. The advent of modern crops may release those working in the fields from much of the tedium of subsistence agriculture, but may also lead to an increase in poverty and in migration into cities (FAO 1996a).

Openness

Decisions on whether biotechnology should be used in a particular context will have to be addressed through an open process in which respect is given to all viewpoints and the structure of the society to which the technology is made available is respected. The Cartagena Protocol (Secretariat of the Convention on Biological Diversity 2000) requires that the public be consulted. Consultation should extend from the design of the regulatory system through individual decisions concerning products. There is an expectation that parties to the Protocol will "promote and facilitate public awareness, education and participation concerning the safe transfer, handling and use of living modified organisms in relation to the conservation and sustainable use of biological diversity, taking also into account risks to human health. In doing so, the Parties shall cooperate, as appropriate, with other States and international bodies." In addition, the parties are expected (insofar as their law permits) to "consult the public in the decision-making process regarding living modified organisms and . . . make the results of such decisions available to the public, while respecting confidential information in accordance with Article 21" (Article 23, sections 1a and 2).

Consumer Choice and Rights

Perhaps the simplest way of ensuring that all views are respected is to provide real choice to the consumer. Those who do not wish to eat meat derived from pigs, for example, should be respected in that foods should be labeled to provide them with choice. Some seek simply to avoid GM food; could this be a reason for labeling food or for ensuring that food is not provided that could offend these sensibilities? This issue is particularly important for those who cannot easily purchase food and are being provided with food aid. The inability to purchase food should not strip them of their rights. A balance should be struck between these consumer needs and the expectation of commercial firms that they will be able to operate in a predictable environment (Nuffield Council on Bioethics 1999, para. 1.16).

Exploitation

In terms of control of genetic resources or food resources, two quite different types of exploitation of a position of power may be distinguished:

- Blocking access to products or to technology. Some fear that this will happen on a significant scale if the IPR systems in place are abused. Although this is theoretically conceivable, it goes against the primary interest of owners of such rights, which is to make money out of their ownership by selling the product.

- Dumping unwanted products that have not been properly tested or that are not approved in the industrialized countries.

It is often stated that only 30 crops "feed the world." These are the crops that provide 95 percent of dietary energy (calories) or protein. Wheat, rice, and maize alone provide more than half of the global plant-derived energy intake. These are the crops that have received the most investment in terms of conservation and improvement. A further six crops or commodities—sorghum, millet, potatoes, sweet potatoes, soybeans, and sugar (cane and beet)—bring the total to 75 percent of energy intake. This information is based on data on national food energy supplies aggregated at the global level. When food energy supplies are analyzed at the subregional level, however, a greater number of crops emerge as significant. For example, cassava supplies over half of plant-derived energy in Central Africa, although at a global level the figure is only 1.6 percent. Beans and plantain also emerge as very important staples in particular subregions. These major food crops, as well as others such as groundnuts, pigeon peas, lentils, cowpeas, and yams are the dietary staples of millions of the world's poorer people, though they receive relatively little research and development attention (FAO 1996a). Resource-poor farmers constitute over half the world's farmers and produce 15 to 20 percent of the world's food. These farmers have not benefited as much as others from modem high-yielding varieties. It is estimated that some 1,400 million people, approximately 100 million in Latin America, 300 million in Africa, and 1,000 million in Asia, are now dependent on resource-poor farming systems in marginal environments (FAO 1996a).

Bias against the Poor

One of the issues that has been mentioned on a number of occasions in this report is that the use of modern biotechnology could, if not used in a careful manner that respects the integrity and needs of all, be a force driving increasing inequity. According to the FAO document on ethical issues (FAO 2001, p. 12): "Most societies were once structured so that, even though many people were poor, most had access to sufficient food to ensure their survival. Social, economic and technological changes have since eroded the traditional 'safety nets,' and ties to the land have been weakened or severed, making it difficult or impossible for the poor to grow their own food." Widespread bias against the hungry and the poor is thus viewed as one of the most egregious problems raised by technological advance of any kind. Pressures to recoup the high costs of investment in biotechnology likely create the conditions for additional bias toward solving the problems of the rich.

Animals

There may be intrinsic objections to the use of modern biotechnology when working with animals. It is recognized that particular kinds and degrees of harm should not be inflicted on any animal. When harm is permissible, it needs to be justified and must be outweighed by benefit either to animals in general or to human beings (United Kingdom Ministry of Agriculture, Fisheries and Food 1993). However, such harm must be minimized.

It has been argued that genetic modification of animals is unethical in that it involves humans' playing God. For some whose religious convictions forbid the eating of certain animals, care must be taken to permit them to avoid modified plants and animals into which such animal genes have been placed. Placing human genes in animals or plants may be offensive to some. The Netherlands' Advisory Committee on Ethics and Biotechnology in Animals (1990) has written:

> Traditionally, ethical and juridical systems in Western society are highly human orientated. Insofar as individual animals were valued, the value was derived from the importance of animals to man. . . . The sense of values with regard to animals is shifting. Especially the criticism of the use of animals as experimental animals and of livestock housing has resulted in the recognition that *animals have a value of their own, or an intrinsic value to man*. . . . Animals come to fall under the province of ethics, not in the sense that animals are thought to act morally, but in the sense of deserving moral care. (Emphasis in original)

According to the Royal Society of the United Kingdom (2001): "Application of genetic modification technology to animals can be used in medical research to create models of human disease. Such models help identify disease pathways and allow assessment of new therapies. Analysing gene function is an area in which the use of GM animals is likely to rise significantly, because by modifying a gene, its various roles in different functional systems of the body can be identified." The concept of stewardship is critical for animals, as we perceive them to have feelings but they are not able to fend for themselves.

The use of animals in biotechnology does pose risks. There may be new allergic reactions when humans come into contact with animals or eat them. There may be toxic effects on humans, animals, and other organisms. Changes in behavior may be important, and the bonds between animals within the same family group may be modified by the modification or an animal might have to be taken out of its social context in order to maintain its freedom from disease. It is possible that transgenic animals may be able to transmit to humans and other animals diseases that could not be transmitted before.

Conclusion

The policy choices made by countries that are members of the OECD have been different. The United States chose not to introduce new laws for the products of biotechnology, relying on its existing regulatory structure. The European Union has made the use of modern biotechnology a trigger for regulation, and Canada regulates all novel products. These choices and the resulting concern about the safety of transgenic organisms in the environment have been confusing to those in the least developed countries. Reasons for decisions need to be clear. There is clearly a need to balance benefits to human health and the environment with risks. The risks are often unclear, speculative, and impossible to test. The benefits of the new crops have not yet been fully demonstrated. People need to feel safe and to be assured that their safety, their health, and their beliefs have been taken into account as far as possible before the introduction of new forms of food products. Although it is undoubtedly a useful exercise to observe the arguments and discussions other countries are having or have had when implementing agricultural biotechnology, in the end it is up to each country, whether developed or developing, to assess the benefits and risks to their own culture and environment when deciding the best way to move forward.

Notes

1. The applicable section of the directive reads as follows:

Article 6.2: The following inventions include those that are unpatentable where their exploitation would be contrary to *ordre public* or morality:

- processes for cloning human beings;
- processes for modifying the germ-line genetic identity of human beings;
- uses of human embryos for industrial or commercial purposes;
- processes for modifying the genetic identity of animals which are likely to cause them suffering without any substantial medical benefit to man or animal, and also animals resulting from such processes.

2. Variously and at various times called genetic modification, genetic manipulation, or genetic engineering.

3. The resolution reads as follows:

Biotechnology can be used to promote contrasting aims:

i. to raise agricultural outputs or reduce inputs;
ii. to make luxury products or basic necessities;
iii. to replace chemical herbicides and insecticides or target them more efficiently;
iv. to upgrade pedigree flocks and herds or expand indigenous stock in developed countries;
v. to upgrade plants for industrial use;
vi. to convert grain into biodegradable plastics or into methanol for fuel;
vii. to hasten maturity in livestock or prevent sexual maturation in locusts or in farmed salmon;
viii. to produce more nutritious and better flavoured foods or diagnose tests for bacterial contamination;
ix. to engineer crops for fertile temperature zones or for semi-arid regions;
x. to fight viral epizootic or build up populations of endangered species;

xi. to reduce production of "greenhouse gases" or utilise them in food production;

xii. to clone meat animals for particular markets or form embryo banks to maintain genetic diversity.

4. Some 44 countries in Europe are members of the Council of Europe.

5. As used in this chapter, "'Transgenic organism" is synonymous with "living modified organism" or "genetically modified organism."

6. For example, on chymosin see http://www.ncbe.reading.ac.uk/NCBE/GMFOOD/chymosin.html.

7. At the First Extraordinary Meeting of the Conference of the Parties to the Convention on Biological Diversity, Cartagena, Colombia, and Montreal, Canada, February 22–23, 1999, and January 24–28, 2000).

8. Article 19(3) reads as follows: "The Parties shall consider the need for and modalities of a protocol setting out appropriate procedures, including, in particular, advance informed agreement, in the field of the safe transfer, handling and use of any living modified organism resulting from biotechnology that may have adverse effect on the conservation and sustainable use of biological diversity."

9. In this overview document, "transgenic," "genetically modified," and even "living modified organisms" are used synonymously.

10. The UNEP/GEF Project on the Development of National Biosafety Frameworks; see http://www.unep.ch/biosafety and specifically http://www.unep.ch/biosafety/countries.htm.

11. Article 2(2) of European Union Directive 2001/18 (European Union 2001) provides the following definition: "Genetically modified organism (GMO) means an organism, with the exception of human beings, in which the genetic material has been altered in a way that does not occur naturally by mating and/or natural recombination. . . . [G]enetic modification occurs at least through the use of the techniques listed in Annex I A, part 1." And Annex IA lists these techniques as follows:

(1) recombinant nucleic acid techniques involving the formation of new combinations of genetic material by the insertion of nucleic acid molecules produced by whatever means outside an organism, into any virus, bacterial plasmid or other vector system and their incorporation into a host organism in which they do not naturally occur but in which they are capable of continued propagation;

(2) techniques involving the direct introduction into an organism of heritable material prepared outside the organism including micro-injection, macro-injection and micro-encapsulation;

(3) cell fusion (including protoplast fusion) or hybridisation techniques where live cells with new combinations of heritable genetic material are formed through the fusion of two or more cells by means of methods that do not occur naturally.

12. Article 1 (Objective) of the Cartagena Protocol on Biosafety (http://www.biodiv.org/biosafety/protocol.asp#) states: "In accordance with the precautionary approach contained in Principle 15 of the Rio Declaration on Environment and Development, the objective of this Protocol is to contribute to ensuring an adequate level of protection in the field of the safe transfer, handling and use of living modified organisms resulting from modern biotechnology that may have adverse effects on the conservation and sustainable use of biological diversity, taking also into account risks to human health, and specifically focusing on transboundary movements."

13. Paragraph 29 of the *Advisory Opinion on the Legality of the Threat or Use of Nuclear Weapons* of the International Court of Justice (1996) reads: "The environment is not an abstraction but represents the living space, the quality of life and the very health of human beings, including generations unborn."

14. Regulation no. 258/97 of the European Parliament and of the Council of 27 (1997) concerning novel foods and novel food ingredients. Note that this regulation is about to be substantially modified to take into account the greater public awareness of GM technology since 1997.

15. Article 5.7 reads: "In cases where relevant scientific evidence is insufficient, a Member may provisionally adopt sanitary or phytosanitary measures on the basis of available pertinent information, including that from the relevant international organizations as well as from sanitary or phytosanitary measures applied by other Members. In such circumstances, Members shall seek to obtain the additional information necessary for a more objective assessment of risk and review the sanitary or phytosanitary measure accordingly within a reasonable period of time."

16. For an example of Greenpeace concerns, go to http://archive.greenpeace.org/~geneng/ or http://ge.greenpeace.org/campaigns/intro?campaign%5fid=3942. For an example of Friends of the Earth concerns, go to http://www.foe.org/foodaid/. For an example of Christian Aid concerns, go to http://www.christian-aid.org.uk/indepth/0003bios/biosafet.htm. For an example of British Medical Association concerns, go to http://www.foeeurope.org/GMOs/bma.doc or http://www.saynotogmos.org/bma_statement.htm.

17. The following was reported in the July 1999 issue of *Natural Foods Merchandiser:* "The world's two largest food production companies have decided they no longer will accept genetically modified ingredients for products sold in Europe. Within hours of one another, both Unilever UK and Nestle UK announced a policy change in response to continued demonstrations by European consumers worried about potential consequences of GMO crops."

18. Article 27(1) of the agreement reads: "Patents shall be available for any inventions, whether products or processes, in all fields of technology, provided that they are new, involve an inventive step and are capable of industrial application. Subject to paragraph 4 of Article 65, paragraph 8 of Article 70 and paragraph 3 of this Article, patents shall be available and patent rights enjoyable without discrimination as to the place of invention, the field of technology and whether products are imported or locally produced."

19. The convention was adopted in Paris in 1961 and was revised in 1972, 1978, and 1991. The objective of the convention is the protection of new varieties of plants by an intellectual property right.

20. The neem tree (*Azadirachta indica*) is a tropical evergreen related to mahogany. Native to east India and Burma, it grows in much of southeast Asia and west Africa. The people of India have long revered the neem tree. For centuries millions have used parts of the neem tree for medicinal purposes, for instance, as a general antiseptic against a variety of skin diseases including septic sores, boils, ulcers, and eczema. In particular, neem may be the harbinger of a new generation of "soft" pesticides that will allow people to protect crops in benign ways. The active ingredient isolated from neem, azadirachtin, appears to be responsible for 90 percent of the effect on most pests. It does not kill insects, at least not immediately. Instead it both repels them and disrupts their growth and reproduction.

21. U.S. Patent no. 5,894,079.

22. This provision does not exist in the European Patent Convention.

23. Article 8j says that a nation should, "subject to its national legislation, respect, preserve and maintain knowledge, innovations and practices of indigenous and local communities embody-

ing traditional lifestyles relevant for the conservation and sustainable use of biological diversity and promote their wider application with the approval and involvement of the holders of such knowledge, innovations and practices and encourage the equitable sharing of the benefits arising from the utilization of such knowledge, innovations and practices."

24. See the preamble to the Cartagena Protocol (Secretariat of the Convention on Biological Diversity 2000).

25. Walker's quote is from J. Boyle, "Sold Out," *New York Times,* March 31, 1996, http://www.wcl.american.edu/pub/faculty/boyle/sold_out.htm.

26. Genesis 1:26 reads: "Let man have dominion over the fish of the sea, and over the fowl of the air, and over every living thing that moves upon the earth."

27. For more on issues of GM food and justice, see Nuffield Council on Bioethics 1999, paras. 1.20–1.31.

28. See http://www.unesco.org/human_rights/hrbc.htm.

References

Advisory Committee on Ethics and Biotechnology in Animals. 1990. Untitled document. The Hague, The Netherlands.

Bruntland, G., ed. 1987. *Our common future: The World Commission on Environment and Development.* Oxford, England: Oxford University Press.

Buchdahl, J., and S. Hare. 2000. Bruntland Report. In *Encyclopaedia of the atmospheric environment.* Available online only at http://www.ace.mmu.ac.uk/eae/Sustainability/Older/Brundtland_Report.php.

Burton, A., ed. 1999. *Selling suicide.* London: Christian Aid.

Commission of the European Communities. 2000. *Communication from the commission on the precautionary principle,* 1. Brussels. February 2.

Comstock, G. 2000. Agricultural bioethics. In *Routledge encyclopedia of philosophy,* ed. Edward Craig. London and New York: Routledge.

Council of Europe, Parliamentary Assembly. 1993. Recommendation 1213 on developments in biotechnology and the consequences for agriculture. Text adopted by the Assembly on May 13, 1993 (36th sitting), Strasbourg, France. http://assembly.coe.int/Main.asp?link=http%3A//assembly.coe.int/Documents/AdoptedText/ta93/EREC1213.htm.

———. 1999. Recommendation 1425 on biotechnology and intellectual property. Strasbourg, France.

———. 2000. Recommendation 1468 on biotechnologies. Strasbourg, France.

Eurobarometer. 2000. *Europeans and biotechnology.* Eurobarometer 52.1. Fourth in a series of opinion polls throughout the European Union. Brussels. http://europa.eu.int/comm/research/pdf/eurobarometer-en.pdf.

European Union. 1997. Regulation (EC) 258/97 of the European Parliament and of the Council of 27 January 1997 concerning novel foods and novel food ingredients. *Official Journal of the European Communities* 14.02.1997, L 043, 0001–0007.

———. 1998a. Directive 98/44/EC of the European Parliament and of the Council of 6 July 1998 on the legal protection of biotechnological inventions. *Official Journal of the European Communities* 30.7.98, L 213/13.

———. 1998b. Directive 98/81/EC of 26 October 1998 amending Directive 90/219/EEC on the contained use of genetically modified micro-organisms. *Official Journal of the European Communities* 5.12.1998, L 330, 0013–0031.

———. 2001. Directive 2001/18/EC of the European Parliament and of the Council on the deliberate release into the environment of genetically modified organisms and repealing Council Directive 90/220/EEC. *Official Journal of the European Communities* 17.4.2001, L 106/1.

———. 2003a. Regulation (EC) 1829/2003 of the European Parliament and of the Council of 22 September 2003 on genetically modified food and feed. *Official Journal of the European Communities* 18.1-10.2003, L 268, 0001–0023.

———. 2003b. Regulation (EC) 1830/2003 of the European Parliament and of the Council of 22 September 2003 concerning the traceability and labelling of genetically modified organisms and the traceability of food and feed products produced from genetically modified organisms and amending Directive 2001/18/EC. *Official Journal of the European Union* 18.10.2003, L 268, 0024–0028.

FAO (Food and Agriculture Organization). 1996a. Background documentation prepared for the International Technical Conference on Plant Genetic Resources, Leipzig, Germany, June 17–23. Rome.

———. 1996b. *Rome declaration on world food security.* Rome. http://www.fao.org/docrep/003/w3613e/w3613e00.htm.

———. 2001. *Ethical issues in food and agriculture.* Rome.

FAO and WHO (World Health Organization). 2000. *Safety aspects of genetically modified foods of plant origin.* Report of a joint FAO and WHO consultation. Rome.

Gaskell, G., N. Allum, and S. Stares et al. 2003. *Europeans and biotechnology in 2002.* Eurobarometer 58.0 (2nd edition): A report to the EC Directorate General for Research from the project Life Sciences in European Society. QLG7-CT-1999-00286. March 21.

Government of Canada. 2001a. *A Canadian perspective on the precautionary approach/principle: Discussion document.* Ottawa. http://www.ec.gc.ca/econom/discussion_e.htm.

———. 2001b. *A Canadian perspective on the precautionary approach/principle: Proposed guiding principles.* Ottawa. September. http://www.pco-bcp.gc.ca/raoics-srdc/doc/Precaution/Booklet/booklet-e-allfonts.pdf.

Hallman, W. K., W. Hebden, H. Aquino, C. Cuite, and J. Lang. 2002. Public perceptions of genetically modified foods. In *Americans know not what they eat*. Publication RR-0302-001. Food Policy Institute, Cook College, Rutgers University, New Brunswick, New Jersey, USA. http://www.foodpolicyinstitute.org.

International Court of Justice. 1996. *Advisory opinion on the legality of the threat or use of nuclear weapons.* Part 1, para. 29. July. The Hague, The Netherlands.

Leger Marketing. 2001. *How Canadians perceive genetically modified organisms.* Montreal, Quebec. http://www.legermarketing.com.

Nyathi, N. 2002. Biotech offers new horizons and problems. *Financial Gazette, Zimbabwe*, May 9. http://www.fingaz.co.zw/fingaz/2002/May/May9/1309.shtml.

OECD (Organization for Economic Cooperation and Development). 1993. *Safety evaluation of foods produced by modern biotechnology: Concepts and principles.* Paris.

OECD, Joint Working Party on Trade and Environment. 2002. *Uncertainty and precaution: Implications for trade and the environment.* Document COM/ENV/TD(2000)114/final. Paris. September 5.

Nielsen, L., and B. A. Faber. 2002. *Ethical principles in European regulation of biotechnolgy: Possibilities and pitfalls.* Copenhagen, Denmark: Ministry of Economic and Business Affairs.

Nuffield Council on Bioethics. 1999. *Genetically modified crops: The ethical and social issues.* London. http://www.nuffieldfoundation.org.

Republic of Ireland, Department of the Environment and Local Government. 1999. *Statement by Mr. Noel Dempsey, T.D., Minister for the Environment and Local Government.* Dublin. http://www.environ.ie/DOEI/DOEIPol.nsf/0/058576aeccca70c580256f0f003bc7f0/$FILE/GMO%20Policy%20Cover.pdf.

Royal Society of Canada. 2001. *An expert panel report on the future of food biotechnology prepared by the Royal Society of Canada at the request of Health Canada, Canadian Food Inspection Agency and Environment Canada.* Ottawa.

———. 2002. *Genetically modified plants for food use and human health: An update.* Policy Document 4/02. February. Ottawa.

Royal Society of the United Kingdom. 2002. Genetically modified plants for food use and human health: An update. Policy Document 4/02. February. London.

Royal Society of the United Kingdom, Science Advice Section. 2001. *The use of genetically modified animals.* London.

Secretariat of the Convention on Biological Diversity. 1992. Convention on biological diversity. June 5. http://www.biodiv.org/convention/articles.asp.

———. 2000. Cartagena protocol on biosafety to the convention on biological diversity: Text and annexes. Montreal. http://www.biodiv.org/biosafety/protocol.asp.

South Africa. 1997. Genetically modified organisms act. *Government Gazette,* vol. 383, no. 18029. Cape Town. May.

UNCED (United Nations Conference on Environment and Development). 1992. *Rio declaration on environment and development.* Rio de Janeiro, June 3–14. http//:igc.apc.org/habitat/agenda21/rio-dec.htm.

UNEP (United Nations Environment Program). 1995. *International technical guidelines for safety in biotechnology.* Nairobi. http://www.bmu.de/download/dateien/unep.pdf.

United Kingdom, House of Commons, Science and Technology Committee. 1999. *Scientific advisory system: Genetically modified foods first report.* HC 286-I. London.

United Kingdom, Ministry of Agriculture, Fisheries and Food. 1993. *Ethical implications of emerging technologies in the breeding of farm animals.* London: Her Majesty's Stationery Office.

United Kingdom, Working Party on the Experimental Manipulation of the Genetic Composition of Micro-organisms. 1975. *Report of the Working Party on the Experimental Manipulation of the Genetic Composition of Micro-organisms.* Cmnd 5880. London. January.

U.S. Office of Science and Technology Policy. 1986. Coordinated framework for biotechnology. *Federal Register* 51, pp. 23302–23350. June.

———. 2002. Proposed federal actions to update field test requirements for biotechnology derived plants and to establish early food safety assessments for new proteins produced by such plants. Notice, Office of Science and Technology Policy. Federal Register 67, pp. 50578–50580.

Walker, S. 2001. *The TRIPS agreement, sustainable development and the public interest.* Discussion paper. Gland, Switzerland, and Cambridge, U.K.: IUCN (International Union for the Conservation of Nature); Geneva, Switzerland: CIEL (Centre for International Environmental Law).

Wright, S. 1994. *Molecular politics: Developing American and British regulatory policy for genetic engineering 1972–1982.* Chicago: University of Chicago Press.

WHO (World Health Organization). 1995. *Application of the principles of substantial equivalence to the safety evaluation of foods or food components from plants derived by modern biotechnology.* Report of a WHO workshop. Geneva.

WTO (World Trade Organization). 1994a. *Agreement establishing the World Trade Organization.* Geneva.

———. 1994b. *Agreement on the application of sanitary and phytosanitary measures* (SPS agreement). Geneva. http://www.wto.org/english/tratop_e/sps_e/sps_e.htm.

———. 1994c. *Agreement on trade-related aspects of intellectual property rights* (TRIPS agreement). Geneva.

———. 2001a. *Declaration on the TRIPS agreement and public health.* Adopted on November 14 at the ministerial conference held in Doha, Qatar. Geneva.

———. 2001b. Trading into the future. Geneva.

Chapter 4

Food Safety and Consumer Choice Policy

David Pelletier

Agricultural biotechnology has the potential to help address a wide range of public health, nutritional, agricultural, and environmental problems in developed and developing countries, as described in a wide variety of scientific (NRC 1985), government (Glickman 1999), industry (Council for Biotechnology Information n.d.), and international (Persley and Lantin 2000; FAO 2003) sources. Despite this potential, the commercialization of the first generation of crops based on these technologies has met with concern and protests from consumer and public interest groups (Consumer's Union n.d.), environmental groups (NRDC 2000), and some governments (EC 2000) and scientists (Union of Concerned Scientists n.d.). This conflict has grown to such proportions that it has resulted in the banning or slowing of the commercialization or use of these products in some countries (*Economist* 1999), disrupted the distribution of food aid in drought-stricken southern Africa (*Economist* 2002), reduced U.S. exports of major commodities (*Economist* 2000), affected the value of Wall Street stocks for major agricultural biotechnology companies (*Financial Times* 2000), and become a major issue of contention in the regulation of international trade (*Financial Times* 2003).

Many of the proponents of agricultural biotechnology have suggested that the issue should be resolved through the application of sound science (Prakash and Bruhn 2000) and that it would be unethical to ban the use of or inhibit the potential benefits associated with this technology for addressing serious problems related to public health, nutrition, poverty, and the environment (Leisinger 2000; Pinstrup-Andersen and Schioler 2000). Many of the critics have called into question the adequacy of the scientific knowledge (Wolfenbarger and Phifer 2000; Clark and Lehman 2001) about this technology, questioned its benefits and raised concerns regarding its potential risks (PSRAST 1998), and claimed that regulatory decisions have been based more on politics than on science (Eichenwald, Kolata, and Petersen 2001; Ferrara 2001).

In contrast to the first generation of genetically modified (GM) crops, which have been designed to address production problems, the second-generation crops currently under development will include a much wider range of alterations. One set will involve changes of potential interest to consumers in developed and developing countries, such as changes in the levels and types of specific fatty acids, vitamins, minerals, phytochemicals, and antinutrients (e.g., phytate). A second set of genetic modifications will focus on agronomic, environmental, and nutritional traits relevant specifically in developing countries, such as drought and saline resistance, insect protection, antiviral and antifungal properties, and enhanced iron, zinc, folate, or pro–vitamin A content, among others. In general, the genetic, metabolic, and food compositional changes in these future crops are expected to be more complex than those of the first-generation crops and may pose more complex regulatory questions (FDA 2001; Kuiper et al. 2001).

The purpose of this chapter is to describe the food safety and consumer issues raised by GM foods, with a particular focus on the choices and trade-offs relevant to southern Africa. Although the ultimate focus of the chapter is on the choices and policy trade-offs relevant to southern Africa, it begins with a detailed analysis of how GM foods are regulated in the United States by the Food and Drug Administration (FDA). This is because FDA policies remain the authoritative position of the U.S. government as applied to the United States and, to a large extent, as projected into international and bilateral discussions and negotiations. Therefore, it is important that developing countries become very knowledgeable concerning FDA policies and their scientific, legal, and political bases so that they can engage in those discussions and negotiations on a more equal footing. In addition, an examination of how the scientific, legal, and political considerations were addressed in the U.S. context holds lessons for southern African countries as they ponder the most appropriate institutional and procedural mechanisms for them to use to reach judgments and develop policies of their own.

The second section of this chapter builds on the first by placing the scientific considerations in the southern African context. This section highlights the significant differences between the U.S. and southern African contexts, the even greater scientific uncertainties in the southern African context as compared to the U.S. context, and the implications for research and policy development. The third and final section provides a framework for discussing policy options and trade-offs under conditions of high complexity and uncertainty, such as in GM agriculture.

Sources and Methods

A large body of literature has already emerged concerning the development of agricultural biotechnology policy, most of it in the past five to eight years as a result of the intense controversy. This includes an immense volume of media reports, popular and semipopular books and magazine articles, industry and trade newsletters, reports and commentaries from a wide spectrum of critical and supportive nongovernmental organizations, special issues of or articles in scientific and social scientific journals, and academic books. Most of these sources contain verifiable factual information (e.g., dates of meetings, names of participants, topics discussed, and decisions). However, they also present selective representations and interpretations of scientific knowledge and health and safety risks, reflecting the views of the authors or the organizations.

An important feature of the methods used in this chapter is the heavy reliance on primary sources, such as documents from the *Federal Register*, reports from the National Research Council (NRC, the working arm of the National Academy of Sciences, NAS), and internal memos of the FDA. These sources are used because most of the debate concerning the regulation of GM foods is based on second- and third-hand representations and interpretations of official policy and its justifications as promulgated by the FDA. Such debate is highly prone to perpetuation of the intentional or unintentional distortions and biases of various parties, especially in light of the scientific and legal complexities and ambiguities posed by GM foods. I acknowledge that the use of primary sources and direct quotes is subject to its own methodological pitfalls, such as biased selection of quotes, misinterpretation of quotes, or presentation of them out of context. However, it has the distinct advantage of grounding the subsequent debates about such matters in the "primary data," in keeping with the established norms for deliberation in science and law.

Disclosure

It is appropriate in a chapter of this type to acknowledge and disclose the important role that the author's views and motivations have played in assembling and interpreting the information. During my roughly 20 years as an applied academic I have devoted roughly half my time to food and nutrition problems and policies of developing countries, half to the food and nutrition problems and policies of the United States. My view concerning agricultural biotechnology is that it holds many potential benefits in developed and developing countries, and I am hopeful that ways can be found to realize these benefits while permitting individuals and countries to reduce or manage the uncertainties and risks. I am acutely aware of the extent to which agricultural biotechnology poses a distinct profile of risks and benefits in developed versus developing countries, and my strongest commitments on this issue are to ensure that individual countries can form their own informed judgments and policies.

My first reading of the FDA's 1992 policy in the summer of 2000 suggested that science and politics were poorly articulated and may have been seriously misused in this case, thus giving rise to my further investigations. My subsequent research reinforced these initial impressions. My current research and writing on this issue is motivated in large part by my view that scientific knowledge, good politics, and normative considerations all should occupy prominent and explicit roles in addressing this and similar controversies, and I articulated this view in works published before I developed my current interest in agricultural biotechnology (Pelletier et al. 1999, 2000; Pelletier 2001). As agricultural biotechnology, nutritional fortification, and other efforts to nutritionally alter national and international food supplies move forward, I now see that the ability to integrate scientific knowledge, good politics, and normative considerations into policy development, above all, will require governance mechanisms that are more open, inclusive, transparent, and accountable than they generally are today.

Contextual Differences between the United States and Southern Africa: A Preview

Although the contextual differences between the United States and southern Africa will be addressed in greater detail in the second section of this chapter, it is important to note them explicitly at the outset so that the analysis and critique of the FDA's policies in the first section of the chapter can be interpreted in light of these differences. As shown in Table 4.1, the two contexts differ widely in the nature of their food safety concerns; the prominence of agriculture, food security, and malnutrition in the lives of their people; the nature of their dominant health concerns; and their food regulation systems. This may imply that the potential benefits as well as the potential risks of technological innovation may have a disproportionate impact in the southern African context. For instance, one of the lessons from the Green Revolution was that adoption rates for new technologies often were lower than expected among smallholders because they perceived the potential benefits and risks of new technologies differently than did agricultural scientists, and their heavy reliance on agriculture for survival caused them to be risk-averse.

Populations in the southern African context also rely heavily on a small number of staple foods for the majority of their caloric intake, may consume parts of plants considered inedible in the United States, and employ different methods for

Table 4.1 Contextual differences, United States and southern Africa

Contextual features	United States	Southern Africa
Food safety concerns	Microbiological, chemical, bioterrorism, irradiation, genetic engineering	Microbiological
Types of foods	Highly diverse, processed and prepared or cooked; social and ethnic diversity	A few major commodities; non-Western processing, preparation or cooking, and understanding of what are edible parts
Food insecurity	8 to 10% of population are uncertain about their future access to food	>50% of population have chronic or seasonal food shortages
Causes of food insecurity	Unemployment, low wages, high costs of living, mental or physical disability	Agroclimatic conditions, low productivity, limited economic alternatives
Food quality concerns	Taste, appearance, convenience, healthfulness, emergent social attributes (whether food is sustainable, organic, ethnic, local, GM-free, etc.)	Taste, appearance, processing, storage
Health concerns	Late-onset chronic diseases, obesity, reemergent infectious diseases, aging population	Endemic HIV, infectious diseases, under nutrition and micronutrient malnutrition, young population
Food production and supply	Industrial, national or international distribution, technology-intensive, 2% of population live on farms	Subsistence and local markets, variable technology, majority of population live on farms
Economic base	Large, diversified formal sector and wage economy	Subsistence agriculture, local-scale economies, small formal sector
Food laws and regulations	Extensive, highly developed; high potential for enforcement	Generally limited regulations and enforcement capacity
Drivers of agricultural biotechnology	Industry, government, scientific establishment	Bilateral and international agencies, transnational industry, national scientists and specialists

Source: Compiled by the author.
Note: Some of the entries in this table require modification or elaboration by regional specialists.

food processing, preparation, and cooking, all of which may have a bearing on food safety. Finally, these populations suffer from widespread malnutrition and infectious diseases, including HIV, which may cause or exacerbate food safety problems that would not exist in healthy, well-nourished populations. This may imply that southern African populations may stand to disproportionately experience the benefits and the risks of GM foods, depending on the nature of the modifications and how they interact with the food habits and health or nutritional status of these populations.

The nature of the contextual differences just noted makes it difficult or impossible to render an overall judgment concerning the safety of GM foods in the United States or southern Africa. This is because the outcomes ultimately depend on the nature of the genetic modifications, the metabolic and compositional changes induced by those modifications, and how they interact with various contextual features, as discussed in the next section in the context of the U.S. population.

It is important to note that the terms of reference for this chapter are to examine food safety and consumer choice issues. Those terms of reference do *not* include estimating the potential benefits of GM agriculture for improving food security, nutrition, and health status. This is rather awkward because the examination of policy options and trade-offs very much requires that the risks and the benefits be examined in tandem. Thus the final section of the chapter will suggest a framework for such analysis. But the details will need to be filled in during and after the first of the planned roundtable discussions.

The FDA's Policies for GM Foods

Timeline

Table 4.2 presents a timeline of key events related to the development of agricultural biotechnology policy in general, and the FDA's policy in particular. Policy developments are shown on the left, and a variety of scientific and societal events that shaped or responded to policy development are shown on the right. The policy developments shown are described in detail in the chapter. Due to space constraints the societal developments are not addressed in detail, but these are well described in a number of other sources (Charles 2001; Hart 2002; Winston 2002). This timeline is intended to help the reader follow the policy developments described in later sections of the chapter.

Legal Framework

In 1992 the FDA published "Statement of Policy: Foods Derived from New Plant Varieties" (FDA 1992) in response to numerous requests from industry, academia, and the public to clarify its interpretation of the existing regulatory frameworks as they pertain to plant varieties produced by "the newer methods of genetic modification." The 1992 policy included a review of scientific issues relevant to public health, the regulatory status of GM foods, and labeling, along with guidance to industry concerning how they might meet the FDA's regulatory requirements

Table 4.2 Key events in the development of agricultural biotechnology policy, 1973–2002

Policy developments	Year	Societal and scientific events
Gordon Conference held on the safety of bacterial recombinant DNA (rDNA) experiments	1973	Boyer and Cohen perform gene transfer; Singer and Soll letter appears in *Science*
Asilomar Conference held; voluntary moratorium enacted	1975	
National Institutes of Health forms Recombinant-DNA Advisory Committee (RAC); safety procedures developed	1976	Citizens in Massachusetts and California protest rDNA research
Extensive research and containment procedures address safety questions	1978	rDNA bacterium produces insulin
	1979	Public protests of rDNA research subside; rDNA bacterium produces human growth hormone
Diamond v. *Chakrabarty* permits gene patents	1980	Cloned bacteria produce interferon
	1981	President Reagan initiates deregulation
Gore hearings reveal lack of scientific evidence on environmental safety	1983	Ice-minus bacterium developed; first rDNA transformation of a plant succeeds, with kanamycin resistance gene
Bayh-Dole Act allows university patents; Biotech Working Group formed	1984	Regulatory uncertainty hinders biotech research; National Research Council (NRC) issues promotional report
BSCC (Biotechnology Science Coordinating Committee) formed	1985	NRC issues promotional report
OSTP (Office of Science and Technology Policy) Coordinated Framework adopted (June); Food and Drug Administration (FDA), U.S. Department of Agriculture (USDA), and Environmental Protection Agency (EPA) clarify policies (June); Monsanto executives visit Vice President Bush (late in year)	1986	
Public questions the scope of oversight proposed by agencies; BSCC attempts to resolve oversight	1987	NRC issues promotional report; ice-minus open-air testing begins; National Academy of Science (NAS) white paper defines key principles
Regulatory uncertainties continue	1988	NAS Food and Nutrition Board issues annual symposium report
BSCC unable to reach consensus; OSTP forwards issues to Quayle Council	1989	NRC issues report on introduction of rDNA into the environment; L-tryptophan food supplement kills two dozen people
OSTP proposes Scope of Oversight	1990	

(*continued*)

Table 4.2 (continued)

Policy developments	Year	Societal and scientific events
Quayle Council shapes oversight policy; FDA begins review of FlavrSavr tomato	1991	Gulf War
OSTP issues final Scope of Oversight; FDA issues Statement of Policy; USDA issues proposed rules	1992	Biotech industry rejoices in FDA policy, though some object to political influence in its development; 4,000 citizens request labeling
USDA finalizes its rules	1993	Recombinant bovine somatotropin (rBST) approved by FDA, public protests ensue; Monsanto adopts aggressive strategies under new CEO (Shapiro)
FDA approves FlavrSavr tomato	1994	rBST protests subside
EPA approves *Bacillus thuringiensi* (*Bt*) corn	1995	U.K. and EU approve Roundup Ready soybeans
	1996	GM maize and soybeans commercialized in U.S.; U.K. acknowledges bovine spongiform encephalopathy (BSE) in human deaths
FDA clarifies its consultation policy; USDA eases its regulations; EPA finalizes its regulations	1997	Public protests begin in Europe
EPA approves Starlink maize for animal feed	1998	Pustzai conducts disputed GM potato studies; Bio-Integrity sues FDA
GM foods become U.S.-EU trade issue; FDA holds three public meetings in response to the conflict and receives 35,000 public comments thereafter	1999	European retailers reject GM food; U.K. imposes three-year ban on new GM crops; EU mandates labeling for GM foods; UN biosafety protocol blocked by U.S.+4; Lossey conducts disputed monarch butterfly study; Seattle citizens protest World Trade Organization, transnational corporations, GM foods, etc.; Golden Rice announced and denounced.
NRC issues report on health and environmental safety of pest-protected plants; Bio-Integrity's suit of FDA dismissed	2000	UN biosafety protocol adopted; Starlink maize detected in human food supply
FDA proposes mandatory premarket notification for new GM foods	2001	Starlink removed from human food supply
NRC issues report on environmental effects of transgenic plants; Institute of Medicine initiates report on assessing the unintended health effects of GE food	2002	Chapela and Quist conduct disputed Mexican maize study; southern African drought and GM food aid debates held; NRC issues report on safety of animal biotechnologies; conflict arises between biopharmers and food farmers; pharm-maize contaminates soybean field

Source: Compiled by the author.

before marketing GM foods. The guidance to industry consisted of five decision trees and accompanying text detailing the types of considerations and safety tests that might be performed under various circumstances. The FDA's 1992 policy statement represented an interpretation of how existing regulations were to be applied to GM foods, reflecting the FDA's view that the "newer techniques of plant breeding" (using recombinant DNA or rDNA) did not pose any fundamentally new risks that might require new regulations.

The FDA asserted that it has sufficient authority to regulate GM foods either under the adulteration clause (section 402(a)(1) of the federal Food, Drug and Cosmetic Act),[1] which normally governs whole foods, or under the food additives clause (section 409), which normally governs chemical substances added to foods to achieve an intended effect. GM foods pose a challenge to this binary choice because they are whole foods and have been altered to achieve an intended effect through the "addition" of new segments of DNA and the intended expression product(s). In resolving this issue the FDA had to proceed carefully because the choice would have profound implications for the level and type of premarket testing required, the strictness of the legal safety standard, labeling, the burden of proof placed on various parties, the administrative burden on the FDA, and the pace with which GM foods would enter the marketplace. As noted in the timeline, these policies were being developed throughout the 1980s and the early 1990s, when deregulation was a dominant theme in federal politics and policymaking.

The food additive clause mandates that producers file a food additive petition with the FDA before marketing foods containing an additive, and usually requires that producers perform extensive safety testing to demonstrate that there is "reasonable certainty of no harm" when the additive is used as intended. If successful, this petition results in an affirmative statement from the FDA, in a letter to the producer, stating that the food additive has been approved. All approved food additives must be listed in the ingredients section of the food label. Some added substances can be exempted from the food additive petition process under the GRAS (generally regarded as safe) clause if they have a long history of safe use (e.g., spices, vinegar, and natural flavors) or have been determined to be GRAS in the judgment of qualified experts.

The adulteration clause of the Food, Drug and Cosmetic Act is the authority under which the FDA normally regulates (and recalls) whole foods to guard against microbiological, chemical, or physical contamination. The 1992 policy states: "Section 402(a)(1) of the Act will be applied to any substance that occurs unexpectedly in the food at a level that may be injurious to health.... It is the responsibility of the producer of a new food to evaluate the safety of the food and assure that the safety requirement of section 401(a)(1) is met" (FDA 1992, p. 22990).

Under this clause the FDA typically defines enforcement guidelines known as "action levels" for various contaminants when the identity of those contaminants is known. The prospect of adverse publicity and the threat of legal action normally creates the incentive for industry to adhere to these guidelines and associated good manufacturing practices. However, unlike the food additive clause, the guidelines bear no mandate for premarket testing or for ex ante demonstration that the food meets the higher safety standard of "reasonable certainty of no harm" that applies to food additives. Instead, because these substances occur unexpectedly by definition, a problem with the food typically might be revealed through marketing testing, surveillance, adverse event reports, or outbreaks of illness. In the case of new substances or substances for which action levels have not been defined previously, the food would be considered adulterated "if, by virtue of the added substance, there is a 'reasonable possibility' that consumption of the food will be injurious to health" (FDA 1992, p. 22989).

Thus the food additive clause generally provides greater ex ante assurance of safety for new substances but is more burdensome for producers and for the FDA, while the adulteration clause generally relies upon good manufacturing practices, marketing detection, and recall authority to protect public health.

In its 1992 policy the FDA avoided exclusive use of either the food additive clause or the adulteration clause, and instead sought a type of middle ground. Specifically, (a) there was no mandate for premarket testing or approval; (b) GM foods, as in the case of other whole foods, were presumed to be GRAS unless the details of a specific case suggested otherwise; (c) developers of GM foods, as in the case of developers of other whole foods, were allowed to independently judge whether the new variety was GRAS; (d) developers could voluntarily follow a set of decision trees provided by the FDA to guide their GRAS determination and testing on a case-by-case basis; (e) developers were urged to voluntarily consult with the FDA at the beginning of this process when deciding the protocols they would follow and again at the end to review their findings; and (f) if successful, this process would result not in an affirmative approval letter from the FDA, as in the case of food additives, but rather in a letter that simply reiterated the conclusions the developer had drawn and stated, "FDA has no further questions."

In effect, these guidelines allowed most foods to avoid the higher requirements of the food additive petition process but provided for a greater degree of (voluntary) consultation between the FDA and developers than is the case for non-GE whole foods. In practice, the FDA believes all new varieties marketed to date have gone through the consultation process, but the details on the testing protocols and consultations are not readily available to the public. The FDA's logic and the decision trees achieve this middle ground, in effect, by treating the intended expression products of the transgene (as well as metabolically related nutrients, known toxicants, and known allergens) as the primary focus of premarket assessment and GRAS determination by developers, and treating any unexpected (e.g., pleiotropic or insertional mutagenic) effects of the transformed variety as subject to the marketing adulteration clause (FDLI 1996, p. 94).

As revealed in subsequent sections of this chapter, this approach responded to two powerful considerations: (a) the high-level political mandate to minimize the regulatory interference with this industry and (b) the enormous gaps in scientific knowledge, evidence, and testing methods concerning the unintended consequences of transgenic breeding of food crops, which made it difficult or impossible to produce affirmative evidence of the presence or absence of unintended harmful changes in the new variety.

The profound lack of evidence and testing methods related to the unintended effects of genetic engineering (GE) is a critically important consideration for interpreting the conflicting and contradictory claims related to GM foods. It means that statements from government, industry, and other groups to the effect that "there is no evidence that any of the GM foods currently on the market have caused harm or are unsafe to eat" is primarily a statement about the lack of evidence rather than an affirmative statement regarding safety. It also means that statements from consumer or public interest groups about the dangers or risks of GM foods are primarily statements about the potential for harm rather than about demonstrated harmful effects.

The manner in which the FDA's 1992 policy statement addressed these issues is analyzed in the next section.

Scientific Issues in the FDA's Statement of Policy

In its 1992 policy the FDA notes that a spectrum of techniques exists for genetic modification, including traditional breeding, mutagenesis, somaclonal variation, wide-cross hybridization, protoplast fusion, and the more recently developed rDNA techniques. The FDA notes that all of these techniques have the potential to introduce extraneous genetic material and undesirable traits, and thus they require extensive backcrossing with the parent line to achieve the desired results. Moreover, it asserts that rDNA techniques are superior in this regard: "In theory, essentially any trait whose gene can be identified can be introduced into virtually any plant, and can be introduced without any extraneous material. Since these techniques are more precise [than other forms of genetic modification], they increase the potential for safe, better-characterized and more predictable food" (FDA 1992, p. 22986). This logic forms the basis for the FDA's oft-repeated position that rDNA techniques are simply an extension of genetic modification that has been used by humans for thousands of years, that it creates no fundamentally new risks, and is more precise and predictable than traditional plant breeding.

Although rDNA techniques may be more precise with respect to the genetic material being transferred, this is not the only relevant consideration. Specifically, as the FDA notes, there are scientific reasons why the insertion of the material and the phenotypic effects are not entirely predictable:

> DNA segments introduced using the new techniques insert semi-randomly into the chromosome, frequently in tandem multiple copies, and sometimes in more than one site on the chromosome. Both the number of copies of the gene and its location in the chromosome can affect its level of expression, as well as the expression of other genes in the plant. . . . Additionally, as with other breeding techniques, the phenotypic effects of a trait may not always be completely predictable in the new genetic background of the host. (FDA 1992, p. 22986)

Since this statement was written, these possibilities have come to be referred to as insertional mutagenesis.

The FDA's policy statement notes that a limited number of backcrosses often are performed to enhance the stability of the line and the ability to cross the trait into other lines, but it does not indicate whether this procedure eliminates the unexpected phenotypic effects referred to previously. Moreover, it states that all breeding or genetic modification techniques have the potential to create unexpected effects, but that "plant breeders using well-established practices have successfully identified and eliminated plants that exhibit unexpected, adverse traits prior to commercial use" (FDA 1992, p. 22987).

This statement of reassurance, which appears several places in the statement of policy, does not describe these practices and their efficacy, but some indications are provided in one passage that states: "The established practices that plant breeders employ in selecting and developing new varieties of plants, such as chemical analyses, taste testing and visual analyses, rely primarily on observations of quality, wholesomeness and agronomic characteristics. Historically these practices have proven reliable for ensuring food safety" (FDA 1992, p. 22988). Thus, while stronger methods are available to assess the safety of the intended expression products from the transgene (described below), the statement of policy seems to imply, but does not actually state, that these traditional plant-breeding methods might be sufficient to reduce the likelihood of unintended toxicologic, allergenic, or compositional effects arising from insertional mutagenesis and pleiotropy. We will now examine some excerpts of the policy statement dealing specifically with toxicants and allergens.

Toxicants. One class of potential unintended effects from genetic modification relates to toxicants. The FDA lists several known toxicants found in specific foods (e.g., protease inhibitors in some cereals, lectins and cyanogenic glycosides in some legumes, cucurbiticin in squash and cucumbers, and lathyrogens in chickpeas) and notes that many of these occur at levels that do not cause acute toxicity, while others may cause severe illness or death if foods are not properly prepared.

To guard against inadvertent elevation of known toxicants when creating new varieties, a critical portion of the FDA's guidance to industry states:

> It is not possible to establish a complete list of all toxicants that should be considered for each plant species. In general, the toxicants that are of highest concern are those that have been documented to cause harm in normal or animal diets, or have been found at unsafe levels in some lines or varieties of that species or related species. In many cases, characteristic properties (such as bitter taste associated with alkaloids) are known to accompany elevated levels of specific natural toxicants. If such characteristics provide an assurance that these toxicants have not been elevated to unsafe levels, analytical or toxicological tests may not be necessary. (FDA 1992, p. 22996)

In those cases in which more detailed analytical tests seem warranted, the FDA notes that the interpretation of such tests is complicated by the great variation in levels of naturally occurring toxicants within and between varieties and that great uncertainty exists concerning safe ranges. Thus it states: "In some cases, analytical methods alone may not be available, practical or sufficient for all toxicants whose levels are needed to be assessed. In such situations comparative toxicological tests on new and parental varieties may provide assurance that the new variety is safe. FDA encourages producers of new plant varieties to consult informally with the agency on testing protocols for whole foods when appropriate" (FDA 1992, p. 22996).

As noted, the 1992 policy suggests that the new variety should be compared to parental varieties or to untransformed varieties as a screen for potentially significant changes. The policy states that this is consistent with the concept of substantial equivalence, as developed by the Organization for Economic Cooperation and Development, and with principles discussed in a joint Food and Agriculture Organization–World Health Organization report (FAO-WHO 1991). The FDA's 1992 policy states that comparisons should be made of the following: (a) toxicants and allergens known to occur in the host or donor species, (b) the concentration and bioavailability of important nutrients for which a crop is ordinarily consumed, (c) the safety and nutritional value of newly introduced proteins, and (d) the identity, composition, and nutritional value of modified carbohydrates, fats, or oils.

The concept of substantial equivalence has been further explicated, defended, and critiqued since that time (Millstone et al. 1999; FAO-WHO 2000; IFT 2000) and is one of the subjects currently under study by an NRC committee (NRC n.d.). It suffers from ambiguity concerning what constitutes a meaningful difference in composition, how much statistical power should be present to detect such differences, and whether the new variety should be compared only to the parental variety grown under identical conditions or to the range of values for all untransformed varieties grown under varying conditions. Moreover, it would not, as originally recommended by the FDA, permit identification of unexpected toxicants, allergens, or nutrition-relevant changes because techniques for broad-spectrum profiling gene expression, metabolic intermediaries, and proteins were not available at that time and still are not widely applied for this purpose (Kuiper et al. 2001).

It is noteworthy that a recent Government Accounting Office report (GAO 2002) stated that techniques for broad-spectrum profiling now are becoming available, which would allow for a significantly expanded application of the substantial equivalence concept, including screening for unexpected changes. However, FDA officials and some of the scientists from industry and academia interviewed by the GAO questioned the utility of these techniques because the functional or health consequences of any observed differences may not be known. This logic, if followed in the future, suggests that as more powerful screening methods become available for demonstrating compositional nonequivalence in some plant varieties, the FDA may abandon "compositional substantial equivalence" as the relevant standard in favor of "functional substantial equivalence." It is unclear whether the burden of proof for ascertaining functional equivalence would fall on the manufacturer, on the FDA, on consumer groups, or on the scientific community at large. Nor is it clear whether the new variety would continue to have "presumptive GRAS status" unless or until such adverse consequences were demonstrated.

As reflected in this section and in the decision trees provided by the FDA, the existence of large knowledge gaps, scientific uncertainties, and practical constraints resulted in an FDA policy that requires a high degree of judgment and discretion on the part of producers when deciding how to demonstrate the GRAS status of novel varieties. Since that policy was issued, the FDA has elaborated upon its "evolving approach" to GRAS determinations, with much greater emphasis on independent determinations by producers, much greater reliance on the "common knowledge" component rather than on direct evidence from testing, and a more limited role for the FDA (FDA 1997). As noted, granting discretion to producers was purposely designed into the 1992 policy because GM foods do not fit neatly into either the food additive or the adulteration category. Most of the branches in FDA's decision trees end in the advice that producers "consult FDA," as in the previous excerpt.

This provides flexibility for industry and the FDA but creates problems related to transparency in the regulatory agencies. The NRC (2000) committee stated: "The details of these consultations are not readily available for public scrutiny. If the public wants to obtain documents containing information and data submitted to FDA for consultation, they must request the documents from FDA through the Freedom of Information Act (FOIA). Processing and fulfilling FOIA requests can take a long time" (NRC 2000, p. 175).

In addition to concerns related to known toxicants, the FDA's policy statement notes the potential for creating new toxicants through plant breeding:

> Plants, like other organisms, have metabolic pathways that no longer function due to mutations that occurred during evolution. Products or intermediates of some such pathways may include toxicants. In rare cases, such silent pathways may be activated by mutations, chromosomal rearrangements or new regulatory regions introduced during breeding, and toxicants hitherto not associated with a plant species may thereby be produced. Similarly, toxicants normally produced at low levels in a plant may be produced at high levels in a new variety as a result of such occurrences. (FDA 1992, p. 22987)

The statement of policy goes on to say that the likelihood of this occurring is "considered extremely low in food plants with a long history of use that have never exhibited production of unknown or unexpected toxins" (FDA 1992, p. 22987).

Accordingly, as noted earlier, the decision trees provided as guidance for industry do not require or suggest any methods for screening for such new toxicants. This despite the FDA's clear acknowledgment (quoted earlier) of the scientific reasons why unexpected effects could result not only from reactivation of "silent pathways" but also from pleiotropic effects of the transgene, from insertional mutagenesis, and from differences arising from the functioning of the gene in a new genomic background.

Allergens. The FDA's policy statement says: "FDA's principal concern regarding allergenicity is that proteins transferred from one food source to another, as is possible with rDNA and protoplast fusion techniques, might confer on food from the host plant the allergenic properties of food from the donor plant" (FDA 1992, p. 22987). It notes that while all known allergens are proteins, only a small fraction of the thousands of proteins in the diet have been found to be allergenic, the most common of which are milk, eggs, fish, crustacea, mollusks, tree nuts, wheat, and legumes (notably peanuts and soybeans). In some cases the specific protein in an allergenic food is known, and in other cases it is not yet known. In either case, the FDA states: "Appropriate in vitro and in vivo allergenicity testing may reveal whether the new variety elicits an allergenic response in the potentially sensitive population" (FDA 1992, p. 22987).

In other words, the FDA claims that in vivo and in vitro methods may be capable of testing new varieties for allergenicity in those cases in which the foods or their allergenic proteins are already known, and the FDA's decision trees guide producers in ascertaining which new varieties may warrant such testing. If new varieties are found to be allergenic, such foods could be labeled as such or steps could be taken to eliminate the allergenicity through more refined breeding. However, one of the limitations of allergen testing, even when the identity of the protein is known, is that indirect tests are the only feasible methods, and each has weaknesses. For instance, the amino acid sequences (epitopes) that might signal allergenicity are not known with precision; the in vito digestibility tests may be conducted at nonphysiologic pH levels; tests often are conducted on proteins isolated from bacteria rather than on a food itself, potentially overlooking translational modifications, as in the *Bacillus thuringiensis* (*Bt*) protoxin versus the active endotoxin (NRC 2000); and samples of human sera from sensitive individuals are not sufficiently abundant to permit widespread use of that test (GAO 2002).

Although the FDA's statement of policy is primarily concerned with the eight food types that account for 90 percent of known allergens, it is known that the remaining 10 percent of known allergens are distributed across at least 160 foods (Clydesdale 1996), and many more may exist but not yet have been documented. Allergic reactions are estimated to occur in 1 to 2 percent of adults and in 5 to 8 percent of children (NRC 1998, p. 58). Inasmuch as transgenic techniques are uniquely capable of creating new varieties from vastly different genera of plants (and animals), this widespread distribution of allergens introduces far greater uncertainties and the potential for introducing new allergens, compared to other breeding methods. This would not be a serious concern if producers could test for new allergens. However, as the FDA notes: "[In contrast to the case of known allergens,] a separate issue is whether any new protein in food has the potential to be allergenic to a segment of the population. At this time, FDA is unaware of any practical method to predict or assess the potential for new proteins in food to induce allergenicity and requests comments on this issue" (FDA 1992).

Because of this gap in knowledge, the decision trees offered as guidance to industry do not suggest any direct methods for testing for novel allergens, but instead suggest that producers "consult FDA on protocols for allergenicity testing and/or labeling." It is unclear what further guidance the FDA could provide through those consultations beyond what it provides in the policy statement itself.

Summary

These sections of the FDA's policy statement regarding toxicants and allergens reveal that efforts to ensure the safety of new plant varieties are severely constrained by uncertainties, gaps in knowledge and methods, contextual factors, and practical considerations. These include the following:

- The list of potentially toxic substances in specific varieties of food crops, whose levels may be affected by rDNA insertions, is not known.

- Levels of known toxicants in foods vary widely for genetic and environmental reasons, and the "safe" or acceptable ranges are not known for most of them.

- The sensitivity and specificity of "taste tests" and other indirect tests for predicting the level or safety of toxicants in food is unknown, yet such tests are suggested as a possible screen for toxicants.

- Food allergens are known to be distributed across many foods, far beyond the eight most common ones, and to affect a significant proportion of adults (1 to 2 percent) and children (5 to 8 percent). Inasmuch as rDNA techniques are uniquely capable of transferring genes across vastly different genera and no practical methods exist for testing for new allergens, this appears to create a plausible risk from new allergens but one whose extent and seriousness is largely unknown and for which no tests are presently available.

- The FDA policy assumes that the nature, extent, and frequency of metabolic disruptions, activation, or over-expression of target and nontarget genes resulting from the (semi-random, tandem, and multiple-copy) insertion of new regulatory regions and structural genes is comparable to that from traditional breeding.

Based on this analysis of the 1992 FDA policy, Table 4.3 represents a judgment concerning the effectiveness of the FDA's guidance to industry with respect to various categories of concerns. For the reasons identified earlier, this guidance is likely to be partially effective with respect to known allergens and known toxicants. However, it is ineffective for detecting and preventing exposure to unknown allergens and toxicants and to known allergens and toxicants that arise from various genetic or metabolic disruptions. These patterns are obscured in the policy statement, however, by frequent reference to (a) well-accepted methods that plant breeders use (such as backcrossing and gross morphological inspection) to eliminate undesired traits, (b) the claim that many or all of the unexpected effects are

Table 4.3 The effectiveness of FDA regulations in addressing various categories of concerns in transgenic plants

Categories of concerns	Known[a] toxicants	Unknown toxicants	Known[a] allergens	Unknown allergens
Intended effects of the transgene	E	NE	E	NE
Transcription modification	PE	NE	PE	NE
Pleiotropic[b] effects of the transgene	NE	NE	NE	NE
Insertional effects of the transgene (location, multiple copies)	NE	NE	NE	NE
Effects of regulatory regions (overexpression, activation)	NE	NE	NE	NE
Effects of the genomic background	NE	NE	NE	NE

Source: Author's judgments.
Notes: E = effective; PE = partially effective; NE = not effective.
[a] *Known* refers to knowledge that a given substance or food source is toxic or allergenic; knowledge of effective testing methods; the "safe" or acceptable ranges, if any; and effects of processing methods.
[b] *Pleiotropic* refers to pleiotropy, the common genetic property in which a single gene can influence multiple phenotypic traits and, in this context, may have multiple effects on the chemical composition of plants due to the complexity of metabolic pathways as well as gene-gene interactions.

just as likely with other methods of plant breeding (which has not been demonstrated), (c) "practical constraints" (which actually reflect large gaps in knowledge and methods) that make it difficult or impossible to test for unexpected effects, (d) the implication that the long history of use of the donor and host plants ensures the safety of transgenic varieties, and (e) the suggestion that many or all of these unexpected effects are considered rare. Several of these claims are amenable to testing through scientific procedures, but no such evidence is provided in the policy statement.

It is noteworthy that these uncertainties, knowledge gaps, and potentials for unintended effects were of considerable concern to some of the scientists and scientific administrators who commented on earlier drafts of the 1992 policy statement, as revealed in internal memos made public through a lawsuit brought against the FDA by a coalition of nonprofit organizations (*Alliance for Bio-Integrity v. Shalala* 1998). They also were noted by a committee formed by the NRC to examine the pest-protected crops on the market in the mid- to late 1990s (NRC 2000), which was able to identify only one direct feeding study in a peer-reviewed journal, the disputed and highly controversial study of GM potatoes using rats (Ewen and Pustzai 1999). A search of the food safety literature on Medline, by Domingo (2000), documented a total of 101 food safety papers with the phrase "genetically engineered foods," including 67 papers with the phrase "adverse effects of transgenic foods" and 44 papers with the phrase "toxicity of transgenic foods." Of these, only 8 papers reported findings from original experimental studies of the safety of GE products, all with rodents. Most of the remaining papers offered opinions and commentaries on the safety of GE foods, but without offering supportive data. A similar analysis of research funded by the U.S. Department of Agriculture (USDA) since 1981 confirms a paucity of research on the safety of GE foods (Pelletier 2005).

This paucity of research is in sharp contrast to the rather strong assurances of safety provided by the FDA and proponents of GE foods. It suggests that the phrase "no evidence of harm" so commonly used by the FDA and others is true in the sense that there is little evidence in one direction or the other. This is quite different from the evidentiary standard of "reasonable certainty of no harm" that would have been required if the FDA had chosen to regulate GE foods under the food additive clause of the Food, Drug and Cosmetic Act. As demonstrated in the next section, considering general scientific knowledge concerning insertional mutagenesis, pleiotropy, and other aspects of molecular biology could easily have led the FDA to adopt a more precautionary stance in the 1992 policy statement.

The FDA's 2001 Proposed Rules

As a result of the intense public controversy over GM foods in the late 1990s the FDA held three public meetings in different parts of the United States in 1999, requested written comments on specific questions (and received over 35,000 comments), and subsequently issued proposed rules requiring premarket notification for bioengineered (GM) foods (FDA 2001). The extensive preamble to the proposed rules reveals that the FDA had reconsidered several of its positions articulated in the 1992 policy:

> FDA recognizes that because breeders utilizing rDNA technology can introduce genetic material from a much wider range of sources than previously possible, there is a greater likelihood that the modified food will contain substances that are significantly different from, or are present in food at a significantly higher level than, counterpart substances historically consumed in food. In such circumstances, the new substances may not be GRAS and may require regulation as food additives. (FDA 2001, p. 4709)
>
> FDA believes that in the future, plant breeders will use rDNA techniques to achieve more complicated compositional changes to food, sometimes introducing multiple genes residing on multiple vectors to generate

new metabolic pathways. FDA expects that with the increased introduction of multiple genes, unintended effects may become more common. For example, rice modified to express pro–vitamin A was shown to exhibit increased concentrations of xanthophylls . . . and rice modified to reduce the concentration of a specific protein was found to exhibit an increased concentration of prolamine. (FDA 2001, p. 4710)

There is substantial basis to conclude, however, that there is greater potential for breeders, using rDNA technology, to develop and commercialize foods that are more likely to present legal status issues and thus require greater FDA scrutiny than those developed using traditional or other breeding techniques. (FDA 2001, p. 4711)

Intended changes to the composition or characteristics of the food also could raise safety questions about the food. For example, it is possible that a developer could modify corn so that the corn becomes a significant dietary source of the nutrient folic acid. Folic acid is used to fortify many foods, including breakfast cereals, because of the relationship [with] neural tube defects. However, excess folic acid in the diet can mask the signs of vitamin B12 deficiency. [In addition] it is possible that a modification would be intended to decrease the level of a substance that is considered undesirable, such as the phytate that naturally occurs in soybeans . . . or the fat content of a food. (FDA 2001, p. 4721)

One of the reasons these paragraphs, and the proposed premarket notification in general, are significant is that they overturn two of the fundamental principles expressed in the 1992 policy, namely (a) that there is no difference between GM foods and foods produced through traditional breeding and (b) that the characteristics of the product, not the process, should determine the level of oversight. These principles were used in 1992 to argue that there was no scientific basis for specific regulations for GM foods, but the rules proposed in 2001 would reverse this position. Although the FDA indicates that greater oversight is now required due to the greater scope and complexity of the genetic changes, the 1992 policy statement (and numerous NRC reports in the 1980s) clearly demonstrate that such changes were envisioned prior to the issuance of the 1992 policy. A more plausible reason for FDA's reversal of its earlier position relates to the intense public controversy that arose in the late 1990s.

The rules proposed in 2001 suggest that the FDA could have marshaled a scientific argument for creating specific regulations for GM foods in 1992, but, as described elsewhere (Eichenwald, Kolata, and Petersen 2001), was responding to political pressures from industry and the White House in choosing not to do so at that time. In addition, the previous quotes from the proposed rules highlight the likelihood that nutritionally altered foods may involve more complex genetic and compositional changes than those addressed in the 1992 policy statement. Such changes may require greater oversight, as noted by the FDA, and an enhanced role for nutrition science and professional communities as described in the final section quoted earlier.

The concern over potential unintended compositional changes in GM foods, which was intensified as a result of the public debate in the late 1990s, has generated a small but growing number of studies in the scientific literature directly examining this possibility. Table 4.4 lists all those available at the time of a review conducted in 2001 by Kuiper et al. (2001). These studies confirm that unintended effects can occur as a result of genetic modification, although they do not address whether the frequency and magnitude of differences are different from those of conventional breeding methods or the functional consequences of the observed changes.

Table 4.4 Unintended effects of genetic engineering breeding as of 2001

Host plant	Trait	Unintended effect
Canola	Overexpression of phytoene-synthase	Multiple metabolic changes (tocopherol, chlorophyll, fatty acids, phytoene)
Potato	Expression of yeast invertase	Reduced glycoalkaloid content (−37 to −48%)
	Expression of soybean glycinin	Increased glycoalkaloid content (+16 to +88%)
	Expression of bacterial levansucrase	Adverse tuber tissue perturbations; impaired carbohydrate transport in the phloem
Rice	Expression of soybean glycinin	Increased vitamin B6 content (+50%)
	Expression of pro–vitamin A biosynthetic pathway	Formation of unexpected carotenoid derivatives (beta carotene, lutein, zeaxanthin)
Soybean	Expression of glyphosphate (EPSPS) resistance	Higher lignin content (20%) at normal soil temperatures (20°C); splitting stems and yield reduction (up to 40%) at high soil temperatures (45°C)
Wheat	Expression of glucose oxidase	Phytotoxicity
	Expression of phosphatidyl serine synthase	Necrotic lesions

Source: Modified from Kuiper et al. 2001.
Note: Data are from publicly available reports.

Conclusions Regarding the FDA's GM Foods Policies

This chapter's examination of the FDA's 1992 policy statement on GM foods holds several lessons concerning the roles and uses of science in policy development. These lessons pertain most directly to the first generation of GM foods, but also have relevance to the forthcoming varieties under development.

Many of the potential unintended consequences in the case of GM foods were amenable to scientific investigation to characterize their plausibility and likelihood, frequency, severity, or mitigation, but research on these issues appears to have been sorely neglected, even in the USDA-funded research portfolio. From a science policy perspective, developing the mechanistic knowledge, methods, and tools for investigation of unintended consequences may be a uniquely public-sector responsibility, because the private sector has insufficient incentive to do so. However, the behavior revealed in this case suggests that the prevailing incentives did not favor the investigation of unintended consequences.

The resulting gaps and biases in public research agendas resulted in scientific uncertainties that had a direct and profound impact on the FDA's decision to adopt policies that appeared inadequate to some consumer groups, to some FDA scientists and administrators, to independent scientists, and to governments in other countries. Specifically, this decision

- permitted the default assumption that unintended consequences appear no more likely in GM foods as compared to conventional foods;

- limited the tools and methods available for premarket testing of individual products, and therefore limited the types of tests the FDA could require of developers;

- virtually required the FDA to use only its market authority under the adulteration clause rather than its authority to require premarket testing under the food additive clause; and

- made it possible for the FDA to claim, in the absence of positive evidence of unintended compositional changes and functional consequences, that there was no legal basis for mandating the labeling of GM foods.

Despite the existence of critical gaps and uncertainties in scientific knowledge concerning unintended consequences, key scientific organizations (notably the various committees of the NAS and the NRC, as seen here) displayed overwhelming support for and promotion of biotechnology in general, including GM foods, while devoting little or no concerted effort to investigation of potential food safety risks. Moreover, the NAS and the NRC increasingly have been asked to render scientific judgments on issues with enormous implications for the regulation of GM foods, which has strained their ability to separate the scientific questions from the profound policy implications that have loomed over the members of these committees. This is seen most clearly in the white paper from the five-member committee of the NAS Council (NAS 1987) and the report analyzed in detail in this chapter (NRC 2000).

The FDA's decisions were highly circumscribed by some of its statutes, as well as by high-level political pressure to minimize regulatory interference with this new industry. Within this larger political and legal context, the lack of an empirical database on the actual nature and extent of compositional changes potentially arising from pleiotropic effects or insertional mutagenesis in individual cases, along with the absence of any organized expression of concern from the scientific community, is what permitted the FDA to exercise its discretion in favor of less stringent regulations. In short, while the findings of individual scientists can be rigorous, objective, and neutral, the collective effort and collective knowledge base from the overall scientific enterprise can encompass gross imbalances with respect to risks versus benefits. This, in turn, can have an enormous impact on the policies adopted and, ultimately, on health and nutritional outcomes.

The Southern African Context

While the accounts given earlier in this chapter reveal a number of weaknesses in the FDA's GM food policies for the U.S. population, a number of contextual factors in southern Africa raise additional questions that are not well addressed by the FDA policy. Three of these reviewed in this section relate to cultural differences in food selection and preparation, special issues related to staple foods, and the health and nutritional status of populations in the region.

Cultural Food Selection and Processing Practices

One category of concerns relates to practices for food selection (definitions of edible versus nonedible portions of a plant), processing (storage, soaking, drying), preparation (cooking), and consumption, which can vary widely across cultures and are not well addressed in the FDA's policy statement. For instance, the statement relies heavily on culture-bound terms such as "proper methods of processing," "long history of use," and "normal diets," with an apparent Euro-American referent in mind. This is illustrated in the following quotations:

> Plants are known to produce naturally a number of toxicants and antinutritional factors, such as protease inhibitors, hemolytic agents, and neurotoxins, which often serve the plant as natural defense compounds against pests or pathogens [e.g., protease inhibitors in cereals, lectins in legumes, cyanoglycosides in cassava, glucosinolates in cruciferae, cucurbiticin in squash, lathyrogens in chickpeas]. Many of these toxicants are present in today's foods at levels that do not cause acute toxicity. Others, such as cassava and some legumes, are high enough to cause severe illness and death if the foods are not properly prepared. FDA seeks to assure that new plant varieties do not have significantly higher levels of toxicants than present in other edible varieties of the same species. (FDA 1992, p. 22987)

> This guidance section is primarily designed for the development of new varieties of currently consumed food plants *whose safety has been established by a history of use.* If *exotic species* are used as hosts, testing may be needed to assure the safety and wholesomeness of food. (FDA 1992, p. 22996; emphasis added)

> Processing (cooking) may affect the safety of a substance. This is particularly important in safety assessment of proteins transferred from one food source to another. For example, lectins, which are inactivated by cooking, would raise a safety concern if transferred from a kidney bean, which are eaten cooked, to tomatoes, which may be eaten raw. *The effects of any potential differences in food processing* between the donor and the new plant variety should be carefully considered at each stage in the safety assessment. (FDA 1992, p. 22994; emphasis added)

While some of the italicized sections of these quotes reveal that the FDA is aware of the importance of food processing methods for the safety of conventional and GM foods, its 1992 policy statement does not explore the implications of this for GM foods created in developed countries and exported to developing countries through commercial or food aid channels.

The NRC report (2000) revealed a greater awareness of the cultural differences in food preparation that could affect the safety of novel foods, but did not explore its food safety implications when GM foods are moved across national and cultural boundaries:

> Depending on the protein, a plant modified to express high concentration of inhibitors in edible tissues can cause adverse health effects if the plant is consumed raw, and such a risk can be reduced by designing transgenes that are expressed only in nonedible plant parts. (NRC 2000, p. 57)

The "edible" portion of a plant varies with the species and the consumer in question. In the human diet, the part eaten can also vary with the cultural background of the consumer. (p. 72)

In summary, the FDA policy statement reveals a predominant focus on factors that may affect the safety of GM foods when consumed by the U.S. population, and it does not appear that those writing it considered the wide variety of food habits and practices in other cultural contexts that could have a bearing on the safety of the same food. *This suggests that blanket assurances concerning the safety of new varieties may not be appropriate in some cases in which they have been offered, without detailed knowledge of the contextual factors that may affect the safety of a specific product in a distinctive context.* This may not be a major factor at the present time because of the limited number of GM crops on the market, but may become a very important factor in the future as the variety of GM products increases and they come to be marketed and consumed in diverse countries and cultures. It also is relevant to the development and safety testing of GM varieties within developing countries.

Special Considerations for Staple Foods

Perhaps the most significant "cultural oversight" in FDA's policy is revealed in the section headed "Issues Specific to Animal Feeds," which states: "Unlike a food in the human diet, an animal feed derived from a single plant may constitute a significant portion of the animal diet. For instance, 50 to 75 percent of the diet of most domestic animals consists of field corn. *Therefore, a change in nutrient or toxicant composition that is considered insignificant for human consumption may be a very significant change in the animal diet*" (FDA 1992, p. 22988; emphasis added).

Although this passage claims that "the human diet" does not rely heavily on a single crop, the reality is that the majority of people in developing countries, especially the poor, do subsist on diets with 50 to 75 percent of the calories coming from a single staple food (FAO 1999). In addition, these staple foods in developing countries undergo quite different food processing methods than those used in the United States and other developed countries. It is well known that processing methods and the physiological state of the consumer can greatly affect the stability of potentially allergenic proteins and toxins during processing and after ingestion (Taylor and Lehrer 1996). *The net effect of these differences is that the effective dose of potential allergens (or toxins) to which southern African consumers may be exposed may be many times higher than that assumed for the U.S. population.*

To illustrate the magnitude of the differences between the U.S. diet and diets in southern Africa, it is instructive to examine some of the key conclusions drawn from evaluation of the Starlink maize contamination that occurred in the United States. Starlink maize is one of the *Bt* varieties of genetically modified maize, and in 1998 it was approved by the U.S. Environmental Protection Agency (EPA) for use in animal feed. (The EPA is responsible for reviewing the safety of such products because the transgenic protein (CRY9C) is classified as a plant pesticide.) The product was not approved for human consumption because in the judgment of the EPA (but not that of the company) the extensive tests conducted on the CRY9C protein could not rule out its potential allergenicity. However, in 1999 it was determined (first by a nongovernmental organization (NGO) and subsequently confirmed by government testing) that the human food chain had been inadvertently contaminated with Starlink maize. In the course of extensive investigations, the EPA Science Advisory Panel (consisting of external scientists) concluded that there remained a "medium likelihood" that the CRY9C protein is an allergen, but it had a "low probability to sensitize some individuals" in the United States because of the short duration of exposure, the low concentration of CRY9C in the overall maize supply (due to mixing with other varieties), the processing methods used, and the very low dietary intakes of maize products in the United States (EPA 2000b).

To underscore the latter point, the 95th percentile for dietary intake of whole maize grain (equivalents) in the United States is estimated to be 62 grams per day (EPA 2000a). Even for the segment of the population with the highest level of maize consumption (Hispanics) the 95th percentile is only 88 grams per day. These upper levels of intake are a mere fraction of the intakes common in the southern African region,[2] and the processing methods used in that region are unlikely to denature and degrade the proteins to the same extent as those used in the U.S. context.

The important point about these calculations is *not* that Starlink maize, or the food aid shipments in 2002, were necessarily unsafe for human consumption in the region. Rather the Starlink case is offered as a dramatic example of the need for scientists, policymakers, and NGOs in the region to carefully examine the assumptions made in the safety assessments conducted by the United States in light of specific knowledge of how contextual features of the region differ from those of the United States. This is underscored by statements in a U.S. Department of State fact sheet issued on January 17, 2003, which made no mention of the Starlink episode, the limited methods available for assessing allergenicity, or the potentially dramatic differences in maize consumption levels and processing methods between U.S. populations and those in southern Africa:

> To-date, scientific evidence demonstrates that these commercially available bio-engineered commodities and processed foods are as safe as their

conventional counterparts. The food safety assessments were conducted to evaluate potential risks for the multi-ethnic U.S. population, and the United States is not aware of any reason to suggest that these foods would be unsafe for populations in other countries. . . . While these assessments were conducted to evaluate potential food safety and environmental impacts in the United States, it is expected that the issues are similar in Southern Africa. (U.S. Department of State 2003, p. 2)

Health and Nutritional Status in Southern Africa

An obvious difference between populations in the United States and in the southern African region is that the latter suffer from high levels of infectious disease morbidity, protein-energy and micronutrient malnutrition, and compromised immune systems due to HIV during drought and nondrought periods. A search of the scientific literature did not identify any empirical studies examining whether any of these health and nutritional conditions may affect the safety of GM foods, nor did it identify any systematic exploration of the potential mechanisms by which these conditions may increase or decrease the potential for food safety problems. Taking allergenicity as an example, it is possible that food allergens may more easily pass the mucosal barrier in the gastrointestinal (GI) tract if the GI tract has been compromised by parasites and diarrheal disease, thereby triggering an immune response (IgE) in previously sensitized individuals that may not be seen in healthy populations. On the other hand, individuals with compromised immune status due to HIV may be less likely to exhibit the pronounced IgE immune response that is characteristic in food allergies. Although empirical studies will ultimately be required to examine these issues, it would be valuable to conduct a systematic inventory of the possible or plausible biological mechanisms (or hypotheses) related to interactions between GM foods and the health and nutritional problems found in the southern African region.

The Potential Benefits of GM Agriculture and GM Food

Finally, although widespread morbidity and malnutrition have been presented as important contextual factors that may have a bearing on the safety of GM foods for the people of southern Africa, it is important to recognize that these also are major problems in their own right, which GM agriculture may help to address. Although, as noted, it is not the purpose of this chapter to describe these potential benefits and critically analyze the conditions under which they may be achieved, these clearly are major considerations that must be addressed in evaluating policy options and trade-offs, a subject taken up in the next section.

Policy Options and Trade-offs

A Basic Framework of Science and Values

The pervasive and growing importance of science and the new technologies, and the potentially profound social changes they can engender, has raised fundamental questions about how new technologies should be governed in democratic societies and in a world community that espouses democratic principles. Whereas the dominant pattern in the last century has been to employ scientific institutions, such as scientific advisory committees, to provide guidance based on positive theories, there has been a growing recognition of the need to incorporate broader considerations into the deliberative process based on normative theories. Positive theories seek to explain "what is" and are the usual domain of the natural and social sciences, while normative theories seek to describe the way things ought to be and are the usual domain of the humanities, especially ethics and political philosophy. Normally scholars in these two traditions do not consider how their ideas relate to each other (Brunner and Ascher 1992). However, insights from both traditions are becoming increasingly integrated as regulatory agencies, stakeholders, and communities seek to develop more productive and appropriate methods for regulating the risks and benefits of new technologies (Renn, Webler, and Wiedemann 1995; NRC 1996; Coglianese 1997; Stirling and Mayer 1999; Beierle and Konisky 2000; Fischer 2000; Beierle 2002; Klinke and Renn 2002).

In most cases of new technology, collective (public) decisions must be made in the face of great scientific uncertainty. In addition, the affected individuals differ in their susceptibilities, in their circumstances, and in the values they attach to their autonomy, lifestyles, and potential risks and benefits. The central question is this: what role should science and politics play in relation to these collective or public decisions? From a positive theory perspective, "politics" refers to a wide range of processes that influence how diverse values are currently allocated in society. From a normative theory perspective, "good politics" refers to procedures that citizens would feel are fair and appropriate because they have characteristics such as openness, transparency, inclusiveness, and accountability.

A simplified schema for better understanding these relationships is shown in Figure 4.1, which builds on the cause and effect relationships that are at the core of positive scientific inquiry and rationality. The case of *Bt* maize is chosen for illustration, although conceptually similar diagrams and principles apply to the second-generation GM crops. Panel A depicts a variety of cause and effect relationships, each of which has a certain degree of uncertainty associated with it ($\hat{\varepsilon}$). Within a strictly scientific paradigm each of the relationships shown here, and others not shown, would be of equal interest and vigorously pursued. The strictly scientific

FOOD SAFETY AND CONSUMER CHOICE POLICY 141

Figure 4.1 Cause and effect relationships involved in the introduction of *Bacillus thuringiensis* maize as a food for a human population

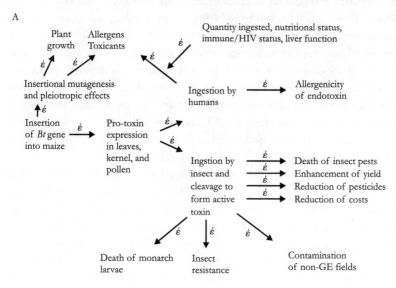

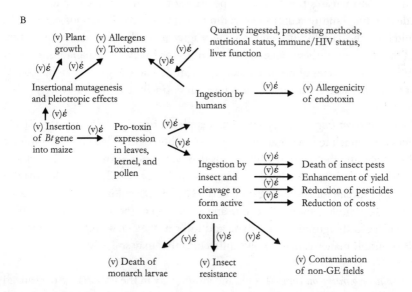

Source: Compiled by the author.

Note: Panel A represents cause and effect relationships as studied by science, with uncertainties (έ); panel B represents these cause and effect relationships with social values included (v); *Bt* = *Bacillus thuringiensis*.

goal would be to test the existence and form of these relationships and understand the mechanisms and contextual factors (effect modifiers) that influence these relationships. This would amount to reducing the uncertainty associated with individual linkages and with the entire causal system.

The relationships among science, politics, and public values can be illustrated very simply through some modifications to this diagram. As shown in panel B, this is accomplished by attaching social values (v) to several elements of this diagram to indicate that different people and groups in society attach different meanings and importance to each of these elements. Although the addition of social values to this diagram appears simple and modest, it has profound implications for the relationship between science and politics in regulatory decisions.

This figure suggests that there are several ways in which GM foods may engender conflicts in social values. These relate to (a) the technology itself, (b) the various outcomes, (c) the uncertainties involved, and (d) boundaries and contexts. In much of the debate concerning GM foods insufficient attention is given to the distinctions among these four categories of values, with the implicit assumption that GM proponents and GM opponents have irreconcilable differences about the value of the technology as a whole. Such a limited view of the normative (or values) dimension of GM increases the chances of polarization, reduces the scope for mutual understanding, and obscures some common interests among various parties that could form the basis for dialogue and policy agreements. For this reason, the nature and implications of these values are explored in the following paragraphs, with an emphasis on the roles of positive theories (scientific knowledge) versus the roles of normative considerations (related to values) in reconciling value differences.

Values regarding technologies. Some people and groups vary in terms of the values they attach to GM as an entire class of technologies. These include intrinsic values regarding the creation of life forms that would not normally exist in nature, as well as extrinsic values related to the possibility that non-GE approaches may be more appropriate for addressing problems related to agriculture, the environment, food security, health, and the structure and ownership of the food system. Scientific knowledge and arguments can shape and inform one's views regarding intrinsic values but ultimately cannot resolve differences that may still exist.

Values regarding outcomes. People and groups vary in the importance they attach to various outcomes, including adverse outcomes (to health, the environment, and agriculture) and beneficial outcomes (to farmers and the environment through reduced losses, costs, and pesticide use). The role of science in such a situation is to estimate, to the best extent possible, the likelihood and magnitude of each of these outcomes and devise ways to enhance the positive ones and minimize the negative ones. However, even with perfect information regarding the various outcomes of using *Bt* corn, there is no scientific method for resolving the value differences among people and groups (Arrow 1963). Moreover, it is inappropriate for scientists or scientific institutions to impose solutions to value-laden issues because, despite their specialized knowledge, "[scientists] remain no better equipped (or mandated) to decide upon profound general questions of values and interests than are any other assemblage of citizens." (Stirling and Mayer 1999, p. 10). The latter point applies equally well to NGOs, despite their claims that they represent the broader "public interest."

The use of market mechanisms is widely recognized as an efficient approach for resolving value differences among individuals, because each person can choose products based on his or her own values. However, the FDA's decision not to impose mandatory labeling of GM foods eliminated this powerful option, and, moreover, some of the outcomes (e.g., environmental ones) involve externalities that are not well addressed through market mechanisms alone. Thus the need remains for collective decisionmaking mechanisms other than science and other than markets to resolve these value differences.

Values regarding uncertainty. People and groups vary in their views of and reactions to uncertainties, and, as shown, uncertainties are pervasive in this causal system. As in the case of outcomes, the appropriate role of scientists, especially those working in public research institutions, is to reduce the degree of uncertainty through research and to improve the methods used to test for allergenicity, toxicity, and other adverse outcomes. As noted, research of this type has been seriously neglected in the GM case, reflecting the lower value placed on unintended consequences by researchers, their institutions, and funding agencies. However, as in the case of outcomes, it is not the role of science or scientists to decide how much and what type of uncertainty should be tolerated by different groups in society. Nor is it the role of science (or of regulators or politicians) to discount or misrepresent these uncertainties in communications with the public, as has been the case with GE.

Insofar as residual uncertainties always will remain, it is notable that three powerful mechanisms exist for managing uncertainty, and especially interindividual differences in risk taking or risk aversion. One efficient mechanism, again, is to permit individual choice in the marketplace. A second is to place legal liability with producers, as the FDA's adulteration clause does in principle. A third risk management method involves insurance markets. As one crude indication of the value Americans place on managing uncertainty, the insurance industry reported sales of $466 billion in the United States in 1998 (WEFA 2000). However, all three of these policy instruments were rendered ineffective in the GM case because the FDA did not impose mandatory labeling and because of the lack of any systematic market surveillance system. This inaction removed the option of consumer choice and made it effectively impossible to establish links between GM foods and any adverse outcomes that might arise. Thus, while labeling and market surveillance might have partially compensated for the scientific uncertainties regarding unintended consequences, the FDA policy precluded even those second-best options.

Values regarding boundaries and context. People and groups differ in the boundaries they place on the breadth and scope of the "causal system" under consideration and on the contextual factors they either include or exclude in their analysis. A forceful example relates to the significant differences in population health and nutrition status that may affect the toxicity or allergenicity of the *Bt* endotoxin and any unintended compositional changes. These contextual or boundary differences were not acknowledged by the FDA or the Department of State.

Finally, as complex as Figure 4.1 and these examples are, they still represent only a small part of the causal system related to GM agriculture. A more complete representation of the causal system would include intellectual property rights; ownership and control of seed stocks and seed companies; long-term effects on ecological systems and on the structure and concentration of agriculture; potential long-term benefits and risks in developing countries; the influence of corporations on politics, regulations, and research funding; the role of the media in promoting the views of GM proponents or critics; public trust or mistrust of government, industry, and scientists and the historic reasons for that; the incentives causing public universities and research centers to do extensive research related to potential benefits and to neglect research related to risks; and so on. Despite the efforts of some GM proponents to limit the boundaries to only those causes and effects shown in Figure 4.1, these broader issues are intimately connected to the GM controversy. Science can play a role in estimating, assessing, and clarifying the nature of these relationships, but it is not the role of science to judge where to set the boundaries.

Science and values in regulatory regimes. Figure 4.2 attempts to integrate these considerations in a way that clarifies the scientific and normative dimensions of the debates over GM foods and other technologies. "Scientific" is defined here in terms of a basic orientation to acquiring knowledge, a broad framing of problems and causal systems, the need for open and accountable social processes such as peer review to verify and challenge accumulating knowledge, and the need to remain open to revision over time. By contrast, "unscientific" approaches are characterized by their lack of openness to challenge, their lack of transparency, their tendency to

Figure 4.2 The relationship between scientific and normative (unscientific) dimensions of regulatory frameworks

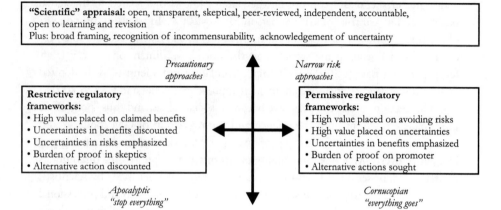

Source: Adapted from Stirling and Mayer 1999.

adopt a narrow view of problems and causal systems, their use of doctrinaire and partisan statements and positions, and their resistance to revision over time. Although many parties, notably the GM proponents inside and outside government, claim to be using "sound science," the evidence reveals that this tends to be backed up more through appeals to institutional authority (the NAS, Biotechnology Science Coordinating Committee (BSCC), FDA, and NIH and the broad scientific community) than by adherence to the characteristics of science shown in this figure.

The figure shows that this scientific dimension can coexist either with permissive or with restrictive regulatory frameworks. These latter concepts are characterized by a basic orientation to technologies and how they should be managed in society, and differences along this dimension also are readily discernible among various parties in the GM debates. It is significant, however, that the four quadrants suggested in this figure are not clearly distinguished in the public discourse, nor are they in the FDA's policy statement and scientific reports from the NAS and other bodies. Instead the overwhelming tendency is to conflate the scientific and normative dimensions and to use the authority of science to support or refute various regulatory approaches (Levidow and Carr 1997). Figure 4.2 demonstrates that a precautionary view, far from being antiscientific, antitechnology, elitist, or immoral, as has been alleged, reflects a broader view of the causal system under consideration and a greater skepticism concerning the state of knowledge related to the actual benefits and actual risks of GE (Auberson-Huang 2002). In addition, proponents of precaution often favor more open, transparent, inclusive, and accountable procedures for deliberating the science and the normative dimensions of GE (Raffensperger and Barrett 2001), while many regulators and scientists in the United States express deep reservations about such approaches (Miller and Conko 2001).

This section reveals the pervasive nature of social values, and thus "politics" in the broadest sense, in both the science and the regulation of GM foods. It also reveals that it is not only industry proponents and activist critics who are engaged in politics over GM foods. Statements or actions that support GE, discount its uncertainties, or set boundaries on the causal system, whether made by scientists, research centers, universities, or scientific institutions (like the NAS and professional societies), are all powerful value statements that explicitly or implicitly promote GM technology even when those parties assert that such statements are purely science-based.

Framing the Policy Goals, Options, and Trade-offs

The previous section suggests that the policy roundtable in southern Africa should help clarify several issues, some of which are covered in the terms of reference for the chapters of this book and some of which are not:

1. what is known about outcomes, that is, the likelihood, frequency, magnitude, and distribution of various outcomes from GM agriculture, based on the best available scientific knowledge and knowledge of local contextual features;

2. the social values attached to each of these outcomes by various groups in society and the policy options for reducing the negative outcomes and enhancing the positive ones;

3. the level of uncertainty associated with various outcomes, the social values attached to that uncertainty, and the policy options for reducing or coping with uncertainty;

4. the relevant boundaries on the issue, which will define which issues are "on the table" for discussion and which are not, and the social values that should guide these decisions; and

5. the nature of the "authorizing institutions" that will be making these decisions as well as the final decisions, the appropriateness of procedures for informing their decisions (e.g., how are social values to be identified—who speaks for whom?), and the methods necessary to ensure openness, inclusiveness, transparency, and accountability in these procedures and decisions.

It is beyond the scope of this chapter to address all of these issues, but the following paragraphs pose contrasting ways to frame the policy questions, outline some distinct policy options, and provide some of the information needed to begin addressing the trade-offs.

There are at least two ways to frame the policy questions related to food safety for this roundtable. They are reflected in the following two sets of questions:

1. Can GM agriculture contribute meaningfully to improving food security and nutrition in southern Africa without creating an unacceptable risk to food safety?

2. What is the relative importance of improving (a) household food security; (b) population nutritional status, especially that of vulnerable groups such as women and children; and (c) morbidity related to food safety? What GM-inclusive policy options and non-GM policy options exist for achieving each of these goals? And what is the full range of potential benefits, risks, and costs associated with each policy option?

Clearly the second set represents a much broader framing of the policy questions and opens the discussion to a much wider set of potentially relevant goals, values, and policy options. While there are some merits to adopting the first question, in that it appears more tractable, the broader goals and social values left "off the table" by that question are problematic and likely will fail to address some of the strongest concerns held by some stakeholders. This section attempts to identify policy options and trade-offs related only to malnutrition, food insecurity, and food safety, recognizing that further options and trade-offs are treated in greater detail in other chapters.[3]

Comparison of problems and uncertainties. Despite the enormous uncertainties implied by the second set of questions, Table 4.5 presents some of the information relevant for addressing those questions. The table suggests that malnutrition and food insecurity are highly prevalent and highly certain problems in the region. By

Table 4.5 Outcomes and uncertainties of genetic modification under GM and non-GM policy options

Uncertainties	Outcomes		
	Malnutrition	Food insecurity	Allergens or toxicants
Onset	Chronic and acute	Chronic and acute	Chronic and acute
Prevalence	10–80% all forms	20–80%	Depends on the nature of the allergen or toxin, individual sensitivity, how widely a commodity is consumed, and the quantitites consumed
Protein-energy	10–50% protein energy	N/A	
Iron deficiency	up to 80% iron deficiency	N/A	
Vitamin A	0–30% vitamin A deficiency	N/A	
Zinc, folate, etc.	5–30% zinc, folate deficiency	N/A	
Probability of occurrence	100%	100%	Uncertain but low
Targets for policy change	Non-GM (current): Food security Diet diversification Supplements Supplemental feeding Fortification Breastfeeding promotion Growth promotion Community-based primary health care Water, sanitation, hygiene Female education Child spacing	Non-GM (current): Agricultural intensification Agricultural diversification Export of agricultural products Nonagricultural income Postharvest technology Market infrastructure Trade Targeted food subsidies Food aid (peace, rule of law, good governance, equity, human rights, international support)	GM (new): Strengthened premarket testing Mandatory standardized profiling methods Context-relevant Export-relevant Public access Public comment Liability incentives Use of test markets Labeling, traceability, segregation Country choice Consumer choice
Marginal impact of GM agriculture on these policy targets	As yet uncertain	As yet uncertain	Not applicable
Issues/questions to be addressed in estimating the potential impact of policy change in these areas	Technical feasibility Efficacy Coverage rates Distribution Acceptance Contextual factors (dietary interactions, parasites, malaria, child feeding practices, etc.)	Technical feasibility Efficacy Variability Adoption rates Distribution Contextual factors (seed markets, performance in local varieties, local agronomy conditions, etc.)	

Source: Compiled by the author.

contrast, problems associated with allergens and toxicants from current and future GM foods are rated here as having a high degree of uncertainty; if they do occur, their prevalence could range from very low to very high (in my judgment).

The basis for this latter judgment is that, in the case of allergens, all known allergens affect only a small proportion of the population and their effects are sufficiently acute and immediate that the offending foods can be quickly identified and avoided. In the case of toxicants, the high end estimate is a worst-case scenario that would occur only if a previously unknown toxicant in a new GM food were toxic to a majority of humans (e.g., lectins in legumes and cyanoglycosides in cassava); were not removed or detoxified through the methods of processing used in a given context; did not affect the taste of the food, the growth and appearance of plants, or other properties that historically have helped to screen out toxic foods; and would escape detection by current premarket testing procedures (which generally focus on known toxicants and have limited ability to screen for unintended and previously unknown toxicants).

GM and non-GM policy options. The second portion of Table 4.5 provides a very brief list of some of the current policy options for addressing malnutrition and food insecurity and for strengthening the safety of GM foods. With respect to malnutrition and food insecurity, the view prior to the advent of GM foods was that these policy options have the demonstrated potential to reduce malnutrition and food insecurity if they are chosen and designed in light of the national and local contexts, are well managed and implemented, and receive the requisite levels of political, institutional, and economic support. In addition, there are some "transboundary" conditions, such as peace, rule of law, good governance, respect for human rights, equity in development, and supportive international institutions that have a powerful bearing on a country's ability to improve the nutrition and food security of its people.

A common concern expressed by critics of GM agriculture is that a technological solution is being advanced for problems that are fundamentally social and political in character, that is, that the more basic policies and changes shown here are required and may be neglected. As suggested in the table, at the present stage of development the marginal impact of GM agriculture might be considered "as yet uncertain." This is due to remaining questions regarding the technical feasibility of developing complex traits such as drought resistance and nutritional improvements and, more important, to questions concerning the efficacy of these changes in light of the diverse national and local contexts in which they might be introduced. It is likely that the ultimate impact of GM agriculture on malnutrition and food insecurity will require continued and even expanded attention to the current policy options. For instance, iron and pro–vitamin A (beta carotene) in plants has very low bioavailability, such that enhanced levels of these nutrients in GM foods may have little or no impact unless the quality of the overall diets also is improved. As another example, enhanced household food security via GM (if achieved) will not reduce child malnutrition unless attention also is given to child health, child care, and child feeding, all of which are constrained by women's health, nutritional status, knowledge, and time demands.

The net effect of these considerations is to suggest that the marginal impact of GM foods on food security and nutrition will depend on simultaneous reduction or elimination of many of the underlying causes of these problems. In addition, these considerations increase the level of uncertainty about the actual effects to be expected from GM foods.

These considerations suggest that a more constructive policy question might be posed as follows: if the success of GM agriculture in improving food security and nutrition requires simultaneous attention to other contextual factors, and if the failure to address these other factors is one of the strong values-based objections to GM agriculture, should the decisions to pursue GM agriculture be tightly linked to firmer commitments to address these contextual factors? Or, put another way, if there are no firm commitments to address the underlying contextual factors, should GM agriculture be pursued?

Strengthening the regulation of GM foods. In the event that GM agriculture is pursued, Table 4.5 suggests a number of ways in which policies could be strengthened to reduce the potential food safety risks of GM foods. These suggestions apply equally to developed and developing countries if problems related to trade are to be avoided. The measures include mandatory (rather than voluntary) pre-market testing of new products, greater standardization of testing methods and decision criteria, and the use of newly emerging broad-spectrum profiling techniques to detect unintended compositional changes (Kuiper et al. 2001). In addition, procedures for developing, testing, labeling, and exporting or importing GM foods should recognize the diverse contexts in which a given GM product may be consumed (and recognize that a food safe in one context may not be safe in another), or the distribution of these foods should be limited to the contexts for which they were intended.

The FDA already has expressed an intent to provide oversight for GM foods developed in other countries and bound for the United States (FDA 2001), but it has not expressed an intent to oversee the export of U.S.-developed GM products to other developed or developing countries. The tacit assumption either is that foods deemed safe in the United States are also safe for other contexts (which can be questioned in light of the contextual factors identified here) or that this oversight is the responsibility of importing countries. In either case, developing countries would need to become very knowledgeable of the testing procedures and results in other countries and be capable of examining them in light of the conditions prevailing in their own contexts.

In addition to issues related to testing and premarket approval, Table 4.5 suggests some procedural and legal changes that would strengthen the incentives for developers to apply rigorous testing methods. These include making the testing protocols and data accessible to the public (already underway at the FDA), providing the opportunity for the public to comment on test results prior to commercialization, and ensuring that the legal liabilities for unintended harm are incentive-appropriate. Mandatory labeling, traceability, and segregation are important for enforcing legal liability, in addition to being important for ensuring consumer choice.

Finally, the use of test markets and monitoring in those markets may be appropriate for some products for several reasons, including (a) the wide variety of products now under consideration and development; (b) the more complex genetic, metabolic, and compositional changes expected in these products; (c) the wide range of contextual factors that may affect their safety; and (d) the increasing knowledge of genetic variation within human populations. This approach would give greater meaning to the claims that "GM foods have been used for years in the United States with no evidence of safety problems" and is consistent with the requirements placed on some producers when controversial or questionable food additives have been introduced in the past.

Summary

Consideration of the relative magnitudes and uncertainties related to the effects of GM agriculture on malnutrition, food security, and food safety suggests that discussions, decisions, and effects related to GM agriculture might be more productive if (a) the development of GM agriculture were tightly linked to firmer commitments to address the underlying causes of these problems and (b) policies were strengthened in relation to the testing, labeling, and marketing of GM foods along the lines suggested here. More fundamentally, this chapter suggests a need for more authentic mechanisms by which governments, stakeholders, and citizens in the southern African region might engage with the scientific and normative dimensions of these issues and develop policies appropriate to the situations, values, and democratic aspirations of the southern African context.

The key food risk concerns identified in this chapter are toxicity and allergenicity. The rDNA techniques used for plant breeding are not simply an accelerated version of traditional plant breeding. There are theoretical reasons to expect a higher degree of unpredictability using these techniques, and this is relevant to the potential for toxicity and allergenicity. Very little empirical experimental work has been done on the safety of GMOs. Policymakers in southern Africa may be tempted to piggyback on the regulatory decisions of developed countries, thinking, "If it is permitted in the United States, we will permit it here." This may not be warranted for two reasons. First, the regulatory framework used in the United States has been based on an imperfect understanding of the science underlying biotechnology, and that regulatory framework is in the process of being modified. Second, the dietary habits in the United States and southern Africa are so different that a product that is "safe" in the U.S. diet is not necessarily safe in the diets of southern Africans.

Notes

1. FDCA, CFR 21 U.S.C. 301 et seq.

2. Assuming an energy intake of 2000 kcal/d, 60 percent of which comes from maize (with an energy content of 350 kcal/100 grams), the typical intake of maize meal in southern Africa would be approximately 340 grams per day.

3. It should be emphasized that the terms of reference for this chapter did not include identifying the policy options to address malnutrition and food insecurity. However, the most common actions to address these problems are presented in Table 4.5, because the analysis of trade-offs with GM food safety concerns could not proceed without them.

References

Alliance for Bio-Integrity v. *Shalala*. 1998. Civil Action No. 98-1300 (CKK). U.S. Court of Appeals for the District of Columbia.

Arrow, K. 1963. *Social choice and individual values.* New Haven: Yale University Press.

Auberson-Huang, L. 2002. The dialogue between precaution and risk. *Nature Biotechnology* 20: 1076–1078.

Beierle, T. C. 2002. The quality of stakeholder-based decisions. *Risk Analysis* 22 (4): 739–749.

Beierle, T. C., and D. M. Konisky. 2000. Values, conflict and trust in participatory environmental planning. *Journal of Policy Analysis and Management* 19 (4): 587–602.

Brunner, R. D., and W. Ascher. 1992. Science and social responsibility. *Policy Sciences* 25: 295–331.

Charles, D. 2001. *Lords of the harvest.* Cambridge, MA: Perseus Publishing.

Clark, E. A., and H. Lehman. 2001. Assessment of GM crops in commercial agriculture. *Journal of Agricultural and Environmental Ethics* 14: 3–28.

Clydesdale, F., ed. 1996. Allergenicity of foods produced by genetic modifications. Special supplement, *Critical Reviews in Food Science and Nutrition* 36: S1–S186.

Coglianese, C. 1997. Assessing consensus: The promise and performance of negotiated rulemaking. *Duke Law Journal* 46 (6): 1255–1349.

Consumer's Union. n.d. http://www.consumersunion.org/food/fdacpi100.htm.

Council for Biotechnology Information. n.d. http://www.whybiotech.com.

Domingo, J. L. 2000. Health risks of GM foods: Many opinions but few data. *Science* 288: 1748–1749.

EC (Commission of the European Communities). 2000. *White paper on food safety.* Brussels. January 12.

Economist. 1999. Who's afraid? June 19.

———. 2000. To plant or not to plant. January 15.

———. 2002. Better dead than GM-fed. September 21.

Eichenwald, K., G. Kolata, and M. Petersen. 2001. Biotechnology food: From lab to a debacle. *New York Times,* January 25.

EPA (Environmental Protection Agency). 2000a. Preliminary evaluation of information contained in the October 25, 2000, submission from Aventis Cropscience. http://www.epa.gov/oscpmont/sap/2000/november/prelim_eval_sub102500.pdf.

———. 2000b. Assessment of scientific information concerning StarLink(TM) corn. Paper presented at the FIFRA (Federal Insecticide, Fungicide and Rodenticide Act) Scientific Advisory Panel meeting, November 28, Rosslyn, VA.

Ewen, S. W. B., and A. Pustzai. 1999. Effects of diets containing genetically modified potatoes expressing *Galanthus nivalis* lectin on rat small intestine. *Lancet* 354 (9187):1353–1354.

FAO (Food and Agriculture Organization). 1999. *State of food insecurity in the world.* Rome.

———. 2003. FAO statement on biotechnology. http://www.fao.org/biotech/stat.asp.

FAO-WHO (World Health Organization). 1991. *Strategies for assessing the safety of foods produced by biotechnology.* Geneva.

———. 2000. *Safety aspects of genetically modified foods of plant origin.* Report of the joint FAO-WHO Expert Consultation on Foods Derived from Biotechnology. Geneva.

FDA (Food and Drug Administration). 1992. Statement of policy: Foods derived from new plant varieties: Notice. *Federal Register* 57: 22984. May 29.

———. 1997. Substances generally regarded as safe: Proposed rule. *Federal Register* 62: 18938. April 17.

———. 2001. Pre-market notice concerning bioengineered foods: Proposed rules. *Federal Register* 66: 4706.

FDLI (Food and Drug Law Institute). 1996. *Basic outlines on food law and regulation: A collective work by top legal and regulatory experts in the food and drug field.* Washington, DC.

Ferrara, J. 2001. Paving the way for biotechnology: Federal regulators and industry PR. In *Redesigning life?* ed. B. Tokar. New York: Zed Books.

Financial Times. 2000. India: Problems with corn sour biotech dreams. November 27.

———. 2003. US ready to declare GM food war. January 10.

Fischer, F. 2000. *Citizens, experts and the environment.* Durham, NC: Duke University Press.

GAO (Government Accounting Office). 2002. *Genetically modified foods: Experts view regimen of safety tests as adequate, but FDA's evaluation process could be enhanced.* Washington, DC. May.

Glickman, D. 1999. New crops, new century, new challenges: How will scientists, farmers and consumers learn to love biotechnology and what happens if they don't? Address to the *National Press Club,* Washington, DC, July 13.

Hart, K. 2002. *Eating in the dark.* New York: Pantheon Books.

IFT (Institute of Food Technologists). 2000. Human food safety evaluation of rDNA biotechnology-derived foods. IFT Expert Report on Biotechnology and Foods. *Food Technology* 54 (9): 15–23.

Klinke, A., and O. Renn. 2002. A new approach to risk evaluation and management: Risk-based, precaution-based, and discourse-based strategies. *Risk Analysis* 22 (6): 1071–93.

Kuiper, H. A., G. A. Kleter, H. P. K. M. Noteborn, and E. J. Kok. 2001. Asssment of the food safety issues related to genetically modified foods. *Plant Journal* 27 (6): 503–528.

Leisinger, K. M. 2000. Ethical challenges of agricultural biotechnology for developing countries. In *Agricultural biotechnology and the poor,* ed. G. J. Persley and M. M. Lantin. Washington, DC: Consultative Group on International Agricultural Research.

Levidow, L., and S. Carr. 1997. How biotechnology regulation sets a risk/ethics boundary. *Agriculture and Human Values* 14: 29–43.

Miller, H., and G. Conko. 2001. Precaution without principle. *Nature Biotechnology* 19 (4): 302–303.

Millstone, E., et al. 1999. Beyond "substantial equivalence." *Nature* 401: 525–26.

NAS (National Academy of Sciences). 1987. *Introduction of recombinant DNA–engineered organisms into the environment: Key issues.* Washington, DC: National Academy Press.

NRC (National Research Council). 1985. *New directions for biosciences research in agriculture: High reward opportunities.* Washington, DC: National Academies Press.

———. 1996. *Understanding risk: Informing decisions in a democratic society.* Washington, DC: National Academy Press.

———. 1998. *Ensuring safe food: From production to consumption.* Washington, DC: National Academy Press.

———. 2000. *Genetically modified pest protected plants: Science and regulation.* Washington, DC: National Academy Press.

———. n.d. Process to identify hazards and assess the unintended effects of genetically engineered foods on human health. http://www4.nationalacademies.org/cp.nsf.

NRDC (Natural Resources Defense Fund). 2000. *NRDC-led coalition calls for USDA to stop environmentally harmful releases of genetically engineered crops.* Washington, DC. http://www.nrdc.org/media/pressreleases/0004262.asp.

Pelletier, D. L. 2001. Research and policy directions. In *Nurition and health in developing countries,* ed. R. D. Semba and M. W. Bloem:. Totawa, NJ: Humana Press.

———. 2005. Science, law and politics in FDA's GE foods policy: Scientific concerns and uncertainties. *Nutrition Review* 63 (6) (in press).

Pelletier, D. L., V. Kraak, C. McCullum, U. Uusitalo, and R. Rich. 1999. The shaping of collective values through deliberative democracy: An empirical study from New York's north country. *Policy Sciences* 32 :103–131.

———. 2000. Values, public policy and community food security. *Agriculture and Human Values* 17 (1): 75–93.

Persley, G. J., and M. M. Lantin, eds. 2000. *Agricultural biotechnology and the poor.* Washington, DC: Consultative Group on International Agricultural Research.

Pinstrup-Andersen, P., and E. Schioler. 2000. *Seeds of contention.* Baltimore: Johns Hopkins University Press.

Prakash, C. S., and C. Bruhn. 2000. Sound science and foods from biotechnology. *San Diego Union Tribune,* June 14. http://www.agbioworld.org/biotech_info/articles/prakash/prakashart/soundscience.html.

PSRAST (Physicians and Scientists for Responsible Application of Science and Technology). 1998. Declaration. http://www.psrast.org/decl.htm.

Raffensperger, C., and K. Barrett. 2001. In defense of the precautionary principle. *Nature Biotechnology* 19: 811–812.

Renn, O., T. Webler, and P. Wiedemann, eds. 1995. Fairness and competence in citizen participation: Evaluating models for environmental discourse. Dordrecht, The Netherlands: Kluwer Academic Publishers.

Stirling, A., and S. Mayer. 1999. *Re-thinking risk: A pilot multi-criteria mapping of a genetically-modified crop in agricultural systems in the UK.* Science and Technology Policy Research, University of Sussex, Sussex, England.

Taylor, S. L., and S. B. Lehrer. 1996. Principles and characteristics of food allergens. *Critical Reviews in Food Science and Nutrition* 36: S91–S118.

Union of Concerned Scientists. n.d. http://www.ucsusa.org/food_and_environment/index.cfm.

U.S. Department of State. 2003. Questions and answers on U.S. food aid donations containing bioengineered crops. Fact sheet, January 17. http://www.state.gov/e/eb/rls/fs/16736.htm.

WEFA (Wharton Economic Forecasting Associates). 2000. *Insurance services: Industry yearbook 2000–2001.* Waltham, MA.

Winston, M. L. 2002. *Travels in the genetically modified zone.* Cambridge, MA: Harvard University Press.

Wolfenbarger, L. L., and P. R. Phifer. 2000. The ecological risks and benefits of genetically engineered plants. *Science* 290: 2088–2093.

Chapter 5

Biosafety Policy

Unesu Ushewokunze-Obatolu

This chapter examines the role of biosafety and its intentions, and the opportunities and challenges that the Southern African Development Community (SADC) region is faced with in connection with research and development in genetic engineering (GE), the importation of GE products, and the movement of such products within and across various SADC countries. It also presents various positions open to the region to explore as it considers the use of biotechnology as one of the tools for agricultural development.

Southern African countries are at different levels of development, including the use of biotechnology. Some countries are receiving assistance from international agencies to develop frameworks for and undertake training in the use of this technology. Recently a number of countries in the region accepted genetically modified (GM) food aid, in most cases before biosafety policies and frameworks were in place. Given the high degree of transboundary movement of goods and people in the region, it is important that decisions by individual countries be open for consideration by neighbors. Further, multinational companies have long been seeking opportunities to introduce biotechnology to develop food and seed industries. A common position is therefore called for to form a basis for biosafety regimes in the interest of food, agriculture, and natural resources for which the SADC already has a policy organ. The success of a biosafety policy framework will depend on country and regional commitment and cooperation, enabling policy instruments, sustainable human and financial support, and enhanced public understanding and awareness of biosafety issues. As a regional group with a development focus based on integration, the SADC is well poised to provide leadership for and guidance to national efforts to develop and enact biosafety policy frameworks.

The Basis for Regulatory Measures in the Life Sciences

Most health problems of humans and animals arise from their close association with the environment, which individuals cannot control but can influence to the detriment of the rest of the population. Human-initiated changes therefore need to be checked to ensure that key public goods continue to be enjoyed without exclusion. The domains of food, human and animal health, and environmental integrity, without reference to biotechnology products, are safeguarded through regulatory measures and policies designed in the public interest. Laws and regulations are developed governing public health, pest control, food and drugs, hazardous substances, agricultural practices, and environmental conservation. Often the aim is to check the exploitative nature of industry and other commercial activities, particularly given the growing need to earn income from new products. Policy, regulatory, and legislative provisions curb private excesses in the interest of society. Such provisions assure consumers and other groups that goods and services produced outside their control will meet certain quality guarantees for their health and welfare. Private businesses that comply may benefit from expanded sales due to enhanced trust.

Potential Risks

Set against the potential benefits biotechnology offers are potential risks. For instance, new organisms could crowd out other organisms, thereby changing ecosystems because of their improved vigor in the environment. GE may alter the internal chemistry of an organism, resulting in undesirable products, some of which could be toxic to other life forms. Some biopesticidal traits conferred through GE could be fatal to susceptible nontarget species. For instance, traits that result in sterility, if applied to insect pests or fishes and passed on though outcrossing, could eliminate certain species, leading to ecological imbalance. Situations could also arise in which mistakes were made, particularly with microbes used in research, whose disposal could lead to massive contamination of water and soil, which would be difficult to rectify and would have detrimental consequences for public health.

Smallholder producers and traders dominate southern African agriculture. In smallholder communities, indigenous genetic resources are often valued for their adaptation to extant conditions and for their medicinal utility. Governments in the region, keen to preserve these traits as public goods, view biotechnology as posing potential barriers to such aims.

Also significant in the region are the risks that biotechnologies may pose to trade, and thus to a range of social welfare concerns. Many governments believe that food imports must not pose risks to human health and the environment. And exports must meet importer's health and environmental requirements.

The International Status of Biosafety

The Cartagena Protocol is a supplement to the Convention on Biological Diversity that seeks to address issues surrounding the safe transfer, handling, and use of living modified organisms (LMOs) resulting from modern biotechnology that may have adverse effects on the conservation and sustainable use of biological diversity in the context of risks to human health, specifically focusing on transboundary movement (CBD Secretariat 2000). Under provisions of the protocol, member countries have an opportunity to assess risks associated with products of GE and indicate their willingness to accept agricultural commodities that include LMOs. Effective implementation of the protocol is linked to the development of national biosafety systems; hence the present efforts to assist countries and regions to develop biosafety regimes. The UNEP-GEF global project on the development of national biosafety frameworks is one such effort (McLean et al. 2002).

The concept of biosafety relates to the World Trade Organization's agreements on sanitary and phytosanitary measures and technical barriers to trade, both of which are about detecting and managing risks for an agricultural trade environment and require risk assessments for decisionmaking support under free trade arrangements. Biosafety provisions also relate to the Codex Alimentarius of the Food and Agriculture Organization and the World Health Organization, which provides voluntary standards on traded food substances.

The general principles of risk or safety assessments were first established by the Organization for Economic Cooperation and Development (OECD 2000). The technical features of practices to assess and manage risks comprise knowledge of the nature of the organism, its products, and distinguishing features of the process by which the product is produced and the environment into which it will be introduced. These are scientifically evaluated on a case-by-case basis once stakeholder concerns have been identified, thereby enabling regulators to identify risks and make recommendations. This implies a requirement of developing new capacities in policy, taking stakeholders on board, and establishing regulatory structures and services. Returns on the development of such systems are maximized if the systems are aligned with international agreements governing movements of genetically modified organisms (GMOs).

The Status of Biosafety in the SADC Region

The biosafety regimes presently in place in the various SADC member countries have to do with conventional pest and disease control in plants, man, and animals; they consist of policies and practices dealing with environmental conservation, food, prophylactics, drugs, cosmetics, and toxic substances. These frameworks require updating or complementing to address products of modern biotechnology.

New products from modern biotechnology still need to be evaluated for their differences from or similarities to known equivalents in terms of their value, safety, and risk. While research has developed modifications for crops grown elsewhere, evaluation of local varieties developed over decades of breeding research is still necessary. Local evaluation also will yield data relevant to local ecosystems. Presently only Malawi, South Africa, and Zimbabwe have biosafety regulations suitable for managing limited or open releases of GMOs. A summary of the status of development and implementation of biosafety systems in the SADC region as of 2001 is given by Mnyulwa (2001). Findings of a southern and eastern African regional workshop on biotechnology (Mswaka, Masimbe, and Mnyulwa 2001) indicated that lack of relevant policies was among the major limitations to the introduction and use of molecular biotechnology. However, an analysis by Cohen and Paarlberg (2002) concludes that nontechnical issues seem to be the deciding factor in the low level of adoption and commercialization of GM technology in developing countries. For Botswana, Namibia, and Zimbabwe, which benefit from a preferential niche market for their beef exports in the European Union, fears of a loss of this prime market contribute to the low level of adoption or reluctance to adopt the technology. A SADC fact-finding mission early in 2003 confirmed that this fear emanates from a European consumer position that is strongly against GM foods for human consumption. While Zimbabwe has biosafety regulations in place, capacity issues may prevent the mainstreaming of testing for genetic modification in meat from beef fed GM feeds, in support of exports.

An approach to setting up biosafety systems is therefore required. Its aim would be to clarify nodes in a decision tree, assess policy alternatives, separate scientific issues from nonscientific ones (McLean et al. 2002), and provide a basis for action plans. A biosafety system will support the already strong seed industry as well as plant and animal genetic resource conservation programs that are in place. Key questions to be addressed include these: Should individual countries develop a national capacity for scientific risk assessment, or should such capacity be developed and coordinated regionally? Should biosafety regulation be centralized in one agency, or should it be distributed among a number of bodies? Should policy harmonization take the form of congruent legislation, or should it merely comprise shared "checklists" of essential elements? When should information about the outcomes of risk assessments be published, and in what forms?

A Biosafety Framework for the SADC

The SADC's 14 member states share objectives for national development based on regional cooperation and integration. The community's Food Agriculture and Natural Resources Sector program aims to meet regional agricultural and natural resources policy objectives revolving around enhanced food security, improved trade, sustainable use of natural resources, and coordinated responses to natural disasters such as drought, floods, and agricultural pests. Mozambique, South Africa, Zambia, and Zimbabwe now enjoy joint actions in managing transfrontier nature parks, emphasizing regional cooperation in the use and conservation of the environment. Through regional cooperation, arrangements for strengthening regional management of transboundary animal diseases and pests supported by quality-accredited testing facilities are under development. Implementation of the Cartagena Protocol will therefore reinforce management of transboundary issues in biosafety from a technical and social standpoint.

McLean et al. (2002) outline a five-point framework to address national needs for countries that are party to the Cartagena Protocol. Table 5.1 represents a preliminary attempt to develop a biosafety framework for the SADC region, building in part on the framework of McLean et al. (2002). This proposed framework is based on a logical process in which an assumed prior position (default or policy position, column 1) is queried through key questions (column 2) about how it will be attained. Depending on how the key questions are answered in the responses (column 3) if answers are necessary, a list of what is to be done (policy instruments, column 4) is stated. Some of the identified policy instruments may need to be further queried, forming a second tier of the decision tree. In the example offered in the table, such policy instruments are marked with an asterisk and brought to column 1 to start the process in the table. The trade-offs in column 5 provide an opportunity to compare exclusive options to enable decisions to be made. A group of stakeholders may treat this exercise more exhaustively in order to maximize the number of questions and trade-off positions suggested. The table shows an example that is likely not exhaustive.

This example complements the global United Nations Environment Program–Global Environment Facility (UNEP-GEF) project on the development of a national biosafety framework (Briggs 2001). It targets the policy environment, including biosafety research agendas and strategies; the resource and knowledge base necessary to assess status and gaps, including capacities and skills; and the development of regulations and implementation of procedures outlined in authority instruments, processes, and procedures for a biosafety system. Having biosafety regimes in place creates a managed environment for the introduction of modern biotechnology, access to products from it, and research and testing that use biotechnology tools. Such a regulated, managed environment creates the confidence required by entrepreneurs and industry, consumers, traders, and those who have responsibility for the technology. It also fosters the development of modern

Table 5.1 Draft of proposed policy development framework for biosafety in the Southern African Development Community

Policy position taken	Key questions	Response	Policy instruments	Trade-offs
No position taken on biosafety	Are measures taken to safeguard the environment and human health?	No	Nil	Indiscriminate and unethical use of biotechnology with threats to human and animal health and the environment
				Social and political dissent with no recourse
				Difficulty of meeting demands made by countries with biosafety policies
				Difficulties with trade partners that affect trade
		Yes	Ad hoc and situational	Actions not well thought out, with negative consequences for food security, technology transfer, resource mobilization, loss of trade opportunities, etc.
				Poor planning and prioritization of actions
				Difficulty in monitoring the status and activities of biotechnology
				Difficulty in coordinating bilateral protocols on biosafety
				Lack of political commitment that undermines the success of situational decisions
Adopt biosafety policy for biotechnology and biosafety under containment	Is there a need for biotechnology under containment? Are authorization channels in place? Are technical capacities available?	Yes	Implement authority, mobilize capacities, and oversee testing and trials	Low public support with absence of direct perceivable benefits to the people (theft of produce, etc.)
			Regionalize trial sites	High investment cost in equipment and personnel, with no prospect of returns by interested parties
		Need identified but no authority or capacity	*Design enabling legislation and regulatory instruments and implement them	Long waiting time for legal drafting or repeals and revision
			Prepare and train personnel to create capacities for inspections and reporting	Numbers of relevant expert scientists low
			*Establish decisionmaking and advisory bodies	
	Do we know what is being safeguarded?	Yes	Support SADC plant genetic resources center with molecular characterization and bioinformatics	High investment cost
			Support conservation at the local level	No system yet for animal genetic resources

BIOSAFETY POLICY   163

	Question	Answer	Actions	Concerns
	Do we know whose interests are being safeguarded?	No	Conduct surveys and field collections and create distribution maps of germ plasm for reference and in situ conservation Establish bioinformatics nodes for local germ plasm (both plant and animal) Establish database of GMOs under test, along with information and decisions pertaining to them Enlist support of the public or farmers who know, use, and are custodians of natural germ plasm Generate information on food safety and human health risks and benefits and provide to public	A slow process due to capacity needs High chances of slow uptake by the public due to highly technical content of subject Perceivable benefits to communities rather small and difficult to grasp
	Are there provisions to deal with cases of noncompliance? Are the provisions enforceable?	No	Design regulations for trials under containment and for reporting of data generated Stipulate liability, redress, and reparation in regulations	Capacity problems with legal personnel Reparation unachievable with some types of gene escape Regulations a disincentive for researchers and investors
	Can results from one country be used in another?		Take measures to ensure adequate capacity Use harmonized procedures in all countries	
Adopt policy for commercialization	Are there any potential benefits and risks from the products or process? Can these risks and benefits be scientifically proven? Are long-term risks assessable?	Yes	*Establish objective measures for benefits and risks for use in informing decisions *Separate scientific and nonscientific risks and benefits for decisionmaking and advice Conduct population epidemiological follow-ups Conduct impact assessment for farming systems and the environment	Some risks and benefits may remain unperceivable Some risks are not measurable using routine laboratory analyses (e.g., some unintended toxins produced in a process despite achieving intended product) Capacity and expertise not sustainable
	Who is affected by this policy?		Inventory stakeholders	Some groups are too diverse and difficult to represent (e.g., farmers: small, medium, large, organic, etc.)
	Are there tracking methods for commercialized (approved) versus unapproved equivalents? Is information about the range of developed GMOs available?		Use reliable standardized test methods Sustain human resource expertise Maintain database of approved GMOs Establish a monitoring system based on transparent information provision by source by means of advance informed agreement principle	Reliance on test protocols developed elsewhere Mutations could occur in local adaptation Cumbersome monitoring system

(continued)

Table 5.1 (continued)

Policy position taken	Key questions	Response	Policy instruments	Trade-offs
*Develop biosafety legislation			Stipulate what is to be monitored (imports, exports, goods in transit, etc.) Identify reliable information source, capturing technology changes and further GM modifications on approved ones	
	Are some existing laws closely associated with biosafety?	Yes	Review them and modify if necessary	Difficulty of modifying several different laws relating to biosafety, some under control by different sectors
		No	Draft new law specific to biosafety	Long time (in years) required, and investment opportunities may be lost to other countries or regions
	Are associated laws in the same sectors?		Define lead sector where related laws are in different sectors	Conflict with other sectors
	Do we know which sector will implement biosafety laws?		Define competent authority for biosafety issues	Differences in capacities in different sectors and biases resulting in advice and decisions
	If in different sectors, do we know how food and agricultural issues will be attended to?		Appoint biosafety focal points for each sector and a lead focal point to coordinate	Difficulty in accessing information from other sectors Difficulty in coordinating cross-sectoral matters Challenge to authority over other sectors Conflict among personnel from different sectors Turn-over of human resource Difficulty in attaining unison at regional level
	Can laws be effectively implemented?	Yes	Design regulatory instruments and quality-assured auditable action plans and procedures	Need to call on external expertise for service audits
	Can laws be effected at the regional level? At the international level?		Harmonize laws at the regional level and with the Cartagena Protocol, the Food and Agriculture Organization–World Health Organization codex, and the World Trade Organization and implement through protocols	Countries may take years to develop laws to be harmonized Need to develop capacity to develop and harmonize laws
	Are the affected members of the public involved?	No	Stipulate use of participatory policy development (social engineering) to maximize ownership	Participatory approaches take time, and there is no guarantee that the outcome will be uniform
	Do we know at what stage the members of the public are to be involved?		If not involved, predetermine points at which members of the public are informed	Informing the public is command controlled, and policy ownership is not ensured

BIOSAFETY POLICY   165

Category	Question	Yes/No	Policy instrument	Issue/Comment
	Do certain groups need to be targeted (e.g., farmers, urban consumers, frontier communities, travelers, etc.)?			
Research and testing	Are policy decisions and regulations guided by scientific evidence?	Yes	Support for priority biosafety research and development (R&D)	Heavy cost of R&D may result in reliance on external evidence sources
			Use of evidence in "decisionmaking	Capacity and cost issues
	Will locally relevant issues be researched for the benefit of the region?	Yes	Support R&D for orphan commodities and local knowledge-based biotechnology for competitive advantage	Local entrepreneurs may not be quick to realize opportunities
			Conduct policy research on the impact of biosafety	Likely to be a long-term action
	Is there capability to conduct tests and trials in regional interests?	No	Do human resource development in biosafety (*risk assessment, research, legislation)	Regional inequalities cause discomfort in training in only a few countries in the region
			Rely on external sources	Relevance problems if focus is not on issues of direct regional interest
	Are resources available for biosafety research?	No	Biosafety research investment position for countries and the region	Competing needs and lack of sustainability for ongoing priorities in research
*Advice and decisionmaking	Do we know how decisions can be made and communicated for implementation?	No	Clarify roles of biosafety focal points, advisory bodies, decisionmakers, regulatory authorities, and reporting structures	Cross-sectoral interests and information leakage
	Do we know who has the final say on decisions made?			Loss of confidentiality by involving the public
	Do we know where information about decisions will be kept?		Clarify roles of expert or advisory committees and the biosafety information hub in communication	
*Risk assessment	Are there local capacities to do risk assessments?	Yes	Appoint institutions or individuals to undertake risk assessments	Empowered regulatory institutions may not have the required expertise
	Do we know what actions will be necessary for products registered elsewhere?		Employ validated auditable procedures based on international norms	Products may not be in use where registered
				Products f interest may not be tested elsewhere

Source: Adapted from McLean et al. 2002 by author.

Note: Asterisks (*) denote policy instruments that need to be further queried, forming a second tier of the decision tree. Policy instruments so marked are brought to column 1 to start the process in the table.

biotechnology within a country, ensuring access to biotechnology products from elsewhere (Persley, Giddings, and Juma 1993). Within the regional context, biosafety regimes are important whether or not products of biotechnology are accepted. Recently a number of countries in the region accepted GM food aid, in most cases before biosafety policies and frameworks were in place. This led to ad hoc decisions ostensibly in the interest of the public and environmental safety. In countries where biosafety regulations were in place, they were invoked for the first time for import commodities, and GM maize could be subjected to strict movement inspections and mandatory milling at ports of entry before distribution.

Land-locked countries may need to use transit routes through neighboring countries to get products to their territories. In addition, certain environmental risks such as those posed by microbes and pollen drift will transcend territorial boundaries, making it necessary to monitor local environments for the presence of unwanted genes. This function will depend on well-managed information systems for coordinated actions.

Biotechnologies are already available in a number of countries of the world, and the SADC region can regulate either to keep them out, in which case it still needs technical capacity and analytical understanding, or to accept them. Multinationals involved in commercial applications with GMOs are applying to test their technologies toward introduction for trials or product development, particularly of seed.

The needs of researchers must also be addressed. Individual countries may wish to accept the technology as a tool only for research and testing or one for research, testing, and commercialization. Either way, biotechnology is unavoidable, and the minimum a country will need will be testing ability that must be accompanied by a biosafety regime for handling a given genetic event, with which reliable diagnosis of GM will be made.

Challenges to Biosafety Policy

Public policies are statements of intent about what is to be done by states or agencies. They are outcomes of interactions between the states or agencies and civil society. Policies are therefore intended to serve the public interest. They are expressed as acts of parliaments or congresses or as regulations that attempt to state in very clear and specific terms what is to be done under various circumstances surrounding an issue. Policies may further be explained for relevance through statutory instruments, guidelines, strategy documents, and action plans. Policymaking in the SADC, as in most developing countries, has tended to be a prescriptive and top-down process rather than one accomplished with public participation. This is due to the low level of literacy that usually obtains, to ignorance about the purpose of policies and regulations, and to the absence of skills in participatory development techniques and the anxiety of administrations eager to bring about changes without committing too much time and financial resources, who therefore implement policies and regulations by force rather than by voluntary cooperation. Although a top-down approach may have worked in developing most past policies, there remains a level of ignorance about the meaning of these policies, as their derivation may not be well understood by the public they are intended to serve. Mandaza (2003) attributes a further difficulty of policymaking in most SADC countries to low levels of interaction across social classes separated by income differentials, which are themselves confounded by race and ethnicity.

Further challenges appear upon recognition that within countries several government ministries are likely to be involved in the policymaking process, each with a different politically motivated position. Ministries of the environment tend to be against biotechnology, normally under pressure from environmental stakeholders and the general conservatism of the United Nations Environment Program, where the environmental agenda is set. Ministries of agriculture (and the national agricultural research institutes that they usually house) and national scientific councils are typically more progressive and would like scientific positions to hold sway. Ministries of trade are conservative and are especially concerned about future prospects for trade with Europe. Ministries of health are conservative and are concerned about implications for human health. Major political logjams can occur. Even when these hurdles have been overcome and legislation has been enacted and is in place, there is typically insufficient capacity in most countries to handle the avalanche of testing that ensues. These capacity constraints are addressed later.

Disparities also exist across countries at different levels of overall economic development—differences that are often determined by and reflective of differences in science and technology policy frameworks. This leads to insecurity in some countries, based on fears of losing revenues and job opportunities and on fears of marginalization and domination of the weak by the strong, which militates against harmonization and collective approaches (*SADC Review* 2001).

Public Involvement

Millions of southern Africans live in poverty in both rural and urban areas. This is in marked contrast to conditions in developed countries where the middle classes dominate, where views about acceptable and expected lifestyles and standards of living are widely held, and, most important, where levels of awareness of public issues are high. The level of public involvement in policymaking is therefore often high, including that relating to biotechnology and biosafety. In contrast, large sections of the public in southern African countries remain totally unaware of biotechnology and biosafety. Those who are aware often hold narrowly defined positions that may be based less on evidence than on politics. For instance, deeply held positions against biotechnologies are often driven by suspicions that countries of the North are using those in the South as dumping grounds for experimental products to provide them with more information before these products can be fully commercialized for use in the North. Instances of public policies supporting exports of toxic waste matter from the North to the South add credence to such positions, which are further strengthened by the increased speed with which information spreads around the globe.

More than 60 percent of the SADC population is engaged in farming, which is closely tied to environmental issues. Food is both formally and informally transported between and within countries. Mechanisms are required to empower citizens by giving them correct understandings of the concepts of the science so that they can articulate and communicate their desires to further the aim of achieving effectiveness and transparency (Cohen 2001). The involvement of the public helps in identifying concerns as well as in seeking ways to address the concerns. It also allows accurate, factual information to be disseminated, thereby dispelling myths spread by rumors (Persley and Doyle 1999).

Public involvement also allows communities to own the process of monitoring their environments for unscrupulous activities and assists regulatory processes through self-policing. Nontechnical issues are crucial to the success of biosafety. Understanding these issues will make it easier for the public to internalize the intentions of regulatory requirements, putting them in a position to assist the often resource-strapped government departments by exercising self-policing on issues related to safeguarding the environment and their health and safety from unwanted or unapproved products.

Of particular concern in this regard are communities who live near frontiers. The frontiers in the SADC are barely 150 years old, established only since the partitioning of Africa. Most are artificial, and the people they divide often belong to the same clans and cultures, so they share heritages, have mutual family connections, and may intermarry. As a result, they often disregard borders, to the detriment of the effectiveness of policies in the countries on either side of these borders.

Another concern related to the safety of the environment is that measures are needed to prevent accidental exposure as goods are transported through foreign territory to reach inland destinations, some of which are land-locked.

The languages of official communication in most of the SADC are foreign, mostly English and to a lesser extent French and Portuguese. Scientific education and laws are written and communicated in these languages. However, more than 70 percent of the region's populations are rural, and in a majority of the countries more than 30 percent are illiterate (SADC Review 2001). Even among the urban dwellers, there are indications that a majority are more comfortable learning concepts and better understand them when using local languages. Local language equivalents still need to be identified for scientific terms.

Capacity

The science itself is relatively new, and only South Africa and Zimbabwe have formal tertiary-level courses. Most biotechnologists have therefore been trained in Europe or North America and are still too few. An even smaller proportion of scientists with training in related disciplines aspire to policymaking positions. The capacity to address public programs in science and technology areas has been affected by high staff turnover due to governments' inability to give staff commensurate rewards and conducive conditions of service. Over the last two decades attrition rates among the highly competent and able-bodied, who comprise the majority in the technical and regulatory professionals, have been rising due to HIV/AIDS as well as the attractions to work under the better-endowed conditions enjoyed at their places of training. High staff turnover affects the ability to sustain policy strategies and actions, critically analyze issues and provide useful advice, and articulate needs, as well as the ability to review and modify the requirements.

Legal services and associated analytical processes are thwarted by a shortage of legal professionals with an understanding of biotechnology. Biotechnology and biosafety know-how may not yet be resident among regulatory service staff. The SADC already lacks institutional capacity at both the national and the regional levels, resulting in a failure to adopt appropriate time-bound performance indicators for its protocol ratification processes and programs (SADC Review 2001). A number of initiatives by regional nongovernmental organizations, including the Southern African Regional Biosafety Initiative and the Regional Agricultural and Environmental Initiative (RAEIN-Africa) aim to address identified scientific capacity and public empowerment, respectively, in biosafety. The UNEP-GEF facility is also assisting with policy formulation and capacity building in some member countries such as Malawi and Namibia.

Financial Resources

Given the poverty levels and increasing fiscal shortfalls of the SADC, traditional funding from member country contributions might fail to meet the requirements.

This factor is likely to compromise concerted actions for biosafety. Policies are more effectively implemented if accompanied by resource allocations. New policies therefore call for additional resources. Investments in public biotechnology and biosafety research could be increased directly by the member states and indirectly through regional collaboration and international partnerships (Cohen 2001) including the private sector as stakeholders. Most donor agencies and investors seem to be increasingly in favor of regional approaches to development.

Interest expressed by multinational companies in registration of their GMO products could be turned into opportunities for resource mobilization for research trials and data accumulation. Issues bordering on conflict of interest will need to be addressed. Local private industries that might benefit from the technology will need to exploit partnerships with the public sector and its agencies to expedite progress in their interest.

Recommendations

The following are my general recommendations related to biosafety policy in the SADC region:

1. The suggested policy framework (Table 5.1) should be considered in order to define appropriate policy alternatives suitable for regional biosafety management toward a ratified protocol.

2. Strategic action plans should be developed to realize the objectives set out to address selected policies.

3. Structures for decisionmaking should be based on benefits and risk assessment, with scientific and other stakeholder concerns used in directing policy instrument design and implementation.

4. Systems to effect regulatory oversight, including quality-controlled and -assured testing for genetic modification, should be developed and introduced.

5. Stakeholder participation in defining biosafety instruments and their objectives should be enhanced.

6. Member countries should be urged to design policies and actions that can be extended into regional and international arrangements.

7. Member countries and the SADC should review their resource base to ensure that they can make effective commitments to allow biosafety processes to begin taking effect sustainably.

8. Member countries and the SADC should review existing biosafety mechanisms, infrastructure, and the human resource base in order to determine which functions can begin immediately and which can be phased in over time according to a schedule.

9. Regional efforts to enhance biosafety research and testing should be promoted to reliably inform regulatory authorities and other regional decision-making structures in order to facilitate movements and trade involving GMOs.

10. Investments should be made in the necessary regulatory, advisory, technical, and legal services in order to identify gaps in biosafety skills and take steps to close those gaps.

11. Investments should be made in systems for the retrieval and exchange of relevant information in order to establish national and regional biosafety information nodes for storage.

12. The legislation and regulatory mechanisms adopted should be sufficiently flexible to account for the dynamism of biotechnology and biosafety and for their rapid development.

References

Briggs, C. 2001. Planning for the development of nearly 100 biosafety frameworks. Paper presented at the international consultation meeting A Framework for Biosafety Implementation—A Tool of Capacity Building, organized by the International Service for National Agricultural Research and Virginia Polytechnic and State University, Washington, DC, July 23–26.

CBD (Convention on Biological Diversity) Secretariat. 2000. *Cartagena protocol on biosafety to the Convention on Biological Diversity: Text and annexes*. Montreal. http://www.biodiv.org/doc/legal/cartagena-protocol-en.pdf.

Cohen, J. I. 2001. Harnessing biotechnology for the poor: Challenges ahead for capacity, safety and public investment. *Journal of Human Development* 2 (2): 239–263.

Cohen, J. I., and R. Paarlberg. 2002. Explaining restricted approval and availability of GM crops in developing countries. *AgBiotechNet,* vol. 4. ABN 097. CAB International, Wallingford, UK. 6 pages.

Mandaza, I. 2003. The problem of schizophrenia in public policy. *The Scrutator.* Southern African Political and Economic Series (SAPES) Trust. Harare, Zimbabwe. February.

McLean, M. A., R. J. Frederick, P. L. Traynor, J. I. Cohen, and J. Komen. 2002. *A conceptual framework for implementing biosafety: Linking policy, capacity and regulation.* Briefing Report 47. International Service for National Agricultural Research, The Hague, The Netherlands.

Mnyulwa, D. 2001. Status of and implementation of biosafety systems in some SADC countries. Report to the Biotechnology Trust of Zimbabwe Board of Trustees. Harare, Zimbabwe.

Mswaka, A. Y., S. Masimbe, and D. Mnyulwa, eds. 2001.The status of biotechnology and biosafety in central, eastern and southern African countries: Defining mechanisms for regional co-operation and networking. *Proceedings of a regional workshop on biotechnology and biosafety.* Speke Resort, Kampala, Uganda. Biotechnology Trust of Zimbabwe, Harare, Zimbabwe.

OECD (Organization for Economic Cooperation and Development). 2000. *Report of the working group on harmonisation of regulatory oversight in biotechnology.* C(2000) 86/ADD2. Paris.

Persley, G. J., and J. J. Doyle. 1999. *Biotechnology for developing country agriculture: Problems and opportunities overview, Focus 2.* Brief 1 of 10. International Food Policy Research Institute, Washington, DC.

Persley, G. J., L. V. Giddings, and C. Juma. 1993. *Biosafety: The safe application of biotechnology in agriculture and the environment.* Research Report No. 5. Intermediary Biotechnology Service. International Service for National Agricultural Research, The Hague, The Netherlands.

SADC Review. 2001. Poverty reduction: A top priority in SADC. www.sadcreview.com/sadc.

Chapter 6

Intellectual Property Rights Policy

Norah Olembo

C hronic hunger persists in most African countries even as crop production reaches peak levels on other continents (Johns Hopkins 2000). In sub-Saharan Africa, more than 600 million people live on small farms measuring no more than a few hectares each. Low productivity due to biotic and abiotic factors is responsible for food insufficiency and malnutrition. The rapid increase in population (nearly 3 percent annually) causes even greater pressure on arable land and is bound to increase the frequency of starvation, for which Africa is so well known.

In Asia, nearly half a century ago the Green Revolution, which used new crop technologies, made increased food production possible. However, Africa has remained sidelined. Today the fastest growing technologies for increased crop production are biotechnologies, whereby inherent crop bioproperties can be manipulated to counter or enhance resistances and tolerances to disease, drought, insect pests, salinity, or nitrogen deficiencies or to improve food value through fortification (Lauderdale 2000; CGIAR 2002; University of California–San Diego and Africa Bio 2002). The annual growth in genetically modified (GM) crops has been more than 10 percent per year since 1996, when GM crops were first planted (IRMA 2002). From 1996 to 2002 the area planted in transgenic crops increased 35-fold globally, from 1.7 to 58.7 million hectares, grown principally by Argentina, Australia, Canada, China, India, South Africa, and the United States. It is noteworthy that of the six leading crops under GM cultivation, five are food crops, with soybeans and maize occupying the largest acreage.

As the focus now turns to critically examining the role of biotechnology in food security for sub-Saharan Africa, key areas have to be analyzed in terms of the different positions of stakeholders and partners. The role of multinational companies and other stakeholders in the application of biotechnology should be defined with respect to biosafety issues and the costs of the technology. Many questions may be asked regarding the trade-offs, that is, the gains and losses of stakeholders, but among the key areas that require attention regarding the use of new biotechnologies to improve food security is the role of intellectual property rights (IPR), not only as it affects the costs of the technology but also as a matter of the gains to be made from reliable policies. This chapter focuses on policy issues concerning IPR in agricultural biotechnology, looking at its positive and negative elements with respect to the positions of stakeholders.

Biotechnology and IPR Issues in Southern Africa: A Need for Policies

The rejection of GM food by authorities in some southern African countries in 2002 and the ensuing confusion of the public comes as no surprise in a region with such little application of GM technology and hardly any policies on it (see Table 6.1). In comparison to high-technology countries, southern Africa, like most of Africa, lags behind in the use of gene technology for food production.

Table 6.1 Status of biosafety regulations and biotechnology policies or laws in eastern and southern Africa, 2004

Country	Status of biosafety regulations	Status of biotechnology policy	
		Policy	Law
Angola	None	None	None
Botswana	None	None	None
Ethiopia	None	Draft	None
Kenya	Guidelines developed by National Biosafety Committee	Draft	Draft
Lesotho	Biosafety Committee established 2001	None	Present
Malawi	National Biosafety Committee established	None	Present
Mauritius	GMO bill for National Biosafety Committee	None	None
Mozambique	None	None	None
Namibia	None	Present	None
Seychelles	None	None	None
South Africa	Present Act 1997	Present	Present
	Legislation enacted	None	None
Swaziland	None	None	None
Tanzania	National Biosafety Committee established	None	None
Uganda	Guidelines or draft regulations written	Draft	None
Zimbabwe	Guidelines established by the Biosafety Board	None	None

Source: Author's compilation.

Table 6.2 Status of laws on intellectual property rights (IPR) in southern Africa, 2004

Country	IPR instruments in place or under way	
	Patent or industrial property law	Plant breeders' rights
Ethiopia	Available	Not available
Kenya	Available	Available—International Union for the Protection of New Varieties of Plants (UPOV) 78
Lesotho	Available	Not available
Malawi	Available	Not available
Mauritius	Available	Not available
Mozambique	Available	Not available
Namibia	Being developed	Not available
Swaziland	Available	Not available
	Available	Available—UPOV 78
Tanzania	Available	Not available
Uganda	Available	Not available
Zambia	Available	Not available
Zimbabwe	Available	Available—national

Source: Author's compilation.

The proceedings of a regional conference on IP and biotechnology in eastern and southern Africa clearly indicate deficiencies in biotechnology policies in most of the 13 countries studied (Kabare and Wekundah 2002). Apart from Kenya, Malawi, Uganda, and Zimbabwe, where national draft policies on biosafety exist, South Africa is the only country with advanced biotechnology policy strategies and the only country in Africa today growing GM crops on a commercial scale (Tables 6.1 and 6.2).

Effective biosafety regulations must have legal backing, that is, they must be supported by an act of a country's parliament or congress. It is for this reason that Kenya has embarked on rigorous discussions to develop a national biotechnology policy and biosafety bill for enactment. In the meantime, existing biosafety guidelines implemented under the National Council of Science and Technology Act are effective in vetting applications for purposes of receiving and handling GM materials as well as carrying out research. For southern African countries, there is an urgent need for implementation of similar processes.

The Importance of IPR Systems

Promoting Innovation
For centuries millions of intellectual property rights have been granted throughout the world under various IP laws of various countries but for similar reasons:

to encourage an inventor (innovator) to disclose his or her invention (innovation) to the public and thereby promote the progress of science and the useful arts. This arrangement may be looked at as a bargain or contract between a government and an inventor whereby the inventor discloses the invention and the government in return provides the inventor with a "monopoly" for a period of time.

This contract is a strong foundation for intellectual property rights, which are governed by laws that create an important government system that provides incentives for inventors or innovators for the development of new technology and ideas for the society.

IPR have revolutionized society technologically, industrially, and thus socio-economically. The doctrine of inventors' disclosing their ideas and governments' granting them monopolies in return has facilitated the enrichment of nations with technological information that is vital not only for promoting the progress of science and the useful arts, but also for the facilitating direct foreign investment through technology transfer.

As a cornerstone of the modern economic policy of any nation and a catalyst for development, IPR have been recognized as important tools for trade and thus have been integrated into global issues like the formation of the World Trade Organization (WTO), to which all the countries of southern Africa are party. The implication of this is that attracting investment in this world's liberalized economy will become harder for countries with weak or ineffective IPR systems. Given that all the southern Africa countries are parties to the WTO, there is a need to develop their IPR systems so that they can participate equitably in the global systems.

Apart from trade facilitation, IP is a rich source of information for the general public on widely diverse research and inventive developments all over the world. IP offices generally are gold mines of such information, which originates in all countries and is stored in databases in national or regional IP offices. Therefore this information is invaluable for industrialization, because detailed descriptions of the inventions can form a basis for manufacturing products. Some of the well-known technologically advanced countries effectively use this information for their industrial development, taking advantage of inventions that have fallen into the public domain. The databases used to store this information can also be used by research institutions in their planning and research and also by government departments for policy development.

In sub-Saharan Africa, IP databases can be accessed at the African Regional Intellectual Property Organization (ARIPO) based in Harare, Zimbabwe, and at the African Intellectual Property Organization. The ARIPO's database holds 30 million patents. Some national offices are currently building up their databases and working toward networking of their offices for easy access under a program supported by the World Intellectual Property Organization (WIPO) based in Geneva. Kenya has 15 million patents in its Documentation Centre, which is accessible to the public.

In spite of the availability of these treasured databases with enormous industrial potential, most members of the public in sub-Saharan Africa hardly ever use them. Extensive publicity and awareness creation is urgently needed to sensitize African governments to the advantages of IP offices as a source of technology for industrial development, including information for production and processing of foodstuffs, pharmaceuticals, chemicals, and equipment.

Regional and International Obligations and the Current Status of IP Knowledge in Southern Africa

Like several other African countries, southern Africa countries have acceded to one or more regional or international laws, treaties, protocols or agreements on intellectual property rights (Table 6.3). These laws obligate member states to protect IPR in their territories. Both the WIPO and the WTO play key roles in the management and enforcement of IP laws internationally.

For example, agropatents are provided for under section 5 of the Agreement on Trade-Related Aspects of Intellectual Property Rights (TRIPS) (WTO 1994). Article 27 of the agreement stipulates that patent protection is available for all inventions in all fields of technology, including agriculture and related sciences. Agroprocesses and agroproducts and their use are patentable, and patent rights are enjoyed without discrimination as to the place of invention, the field of technology, or whether the products are imported or locally manufactured.

Although Article 27(2) of the agreement allows exclusion from patentability of inventions that are contrary to public order or morality, including that regarding the protection of human, animal, or plant life or health or the avoidance of serious prejudice to the environment, Article 27(3)(b) provides that "protection of plant varieties must be done either by patents or by an effective sui generis system or by any combination thereof" (WTO).

Only a few African countries have institutionalized laws for the protection of plant varieties (Table 6.2). The International Union for the Protection of New Varieties of Plants (UPOV) system is viewed with great hostility by most southern African countries with the exception of Kenya and South Africa, which are members of the 1978 UPOV system.

It is not quite understood why southern African countries view the UPOV system with such suspicion, but arguments against it are that the system is excessively monopolistic and protects the breeder to the disadvantage of farmers' rights and indigenous knowledge. This is in relation to clauses in UPOV 91 that prohibit

Table 6.3 Participation of southern African countries in various intellectual property agreements, 2004

Agreement	Participating countries
Madrid Agreement Concerning International Registration of Marks	Algeria, Egypt, Kenya, Lesotho, Liberia, Morocco, Mozambique, Sierra Leone, Sudan, Swaziland, Zambia.
Berne Convention for the Protection of Literary and Artistic Works	Algeria, Benin, Botswana, Burkina Faso, Cameroon, Central African Republic, Chad, Congo, Côte d'Ivoire, Democratic Republic of Congo, Djibouti, Egypt, Gabon, The Gambia, Ghana, Guinea, Guinea Bissau, Kenya, Lesotho, Liberia, Libya, Madagascar, Malawi, Mali, Mauritania, Mauritius, Morocco, Namibia, Niger, Nigeria, Rwanda, Senegal, Sierra Leone, South Africa, Sudan, Swaziland, Tanzania, Togo, Tunisia, Zambia, Zimbabwe
Nice Agreement Concerning the International Classification of Goods and Services for the Purposes of Registration of Marks	Algeria, Benin, Guinea, Malawi, Morocco, Mozambique, Tunisia, Tanzania
Paris Union	Algeria, Benin, Botswana, Burkina Faso, Burundi, Cameroon, Central African Republic, Chad, Congo, Côte d'Ivoire, Democratic Republic of Congo, Djibouti, Egypt, Equatorial Guinea, Gabon, Gambia, Guinea, Guinea Bissau, Kenya, Lesotho, Madagascar, Malawi, Mali, Mauritania, Mauritius, Morocco, Mozambique, Niger, Rwanda, Senegal, Sierra Leone, South Africa, Swaziland, Tanzania, Togo, Tunisia, Uganda, Zambia, Zimbabwe
Hague Agreement Concerning the International Deposit of Industrial Designs	Benin, Côte d'Ivoire, Egypt, Morocco, Senegal, Tunisia
Patent Cooperation Treaty	Algeria, Benin, Burkina Faso, Cameroon, Central African Republic, Chad, Congo, Côte d'Ivoire, Equatorial Guinea, Guinea Bissau, Kenya, Lesotho, Madagascar, Malawi, Niger, Senegal, Sierra Leone, South Africa, Sudan, Swaziland, Tanzania, Togo, Tunisia, Uganda, Zambia, Zimbabwe

Source: Author's compilation.

on-farm sale by the farmer and the sharing of seeds. However, countries that have embraced UPOV 78, such as Kenya, see its advantage as stimulating trade in horticulture, in which access to quality seed and horticultural material such as flowers and vegetables facilitates global trade in these commodities. But perhaps the most significant impact of a plant protection system is its stimulation of research in agricultural productivity.

In order to address issues of farmers' rights and indigenous knowledge, in 2002 the Organization of African Unity published the *African Model Law* for protection of the rights of local communities, farmers, and breeders and for the regulation of access to biological resources (Ekpere 2000). The document is set out as a model for use by African countries that wish to develop their own national laws. However, to date no such laws have been enacted.

Although southern African countries have acceded to one or more regional or international laws (treaties, protocols, or agreements) on IPR, there is a lack of clear-cut policies on IPR in most countries of the region. Formulation of policy and legal frameworks is complicated by the society's lack of appreciation of the role of IPR in development. In recognition of the foregoing, governments of the region need to devote resources to the development of mechanisms for the management of IPR within their territories.

Controversies over IPR in Biotechnology in Southern Africa

IP protection of agrobiotechnology has caused a storm in SADC society. Most of the controversy centers on the threat to food security (Kuyek 2002; Friends of the Earth International 2003; Hivos and Friends of the Earth International 2003). Arguments against IPR are that they confer monopolistic status, placing needed products beyond the reach of poor countries. Fears abound that patents are restrictive and threaten the freedom of farmers to access seed. Ethical questions are asked as to whether private companies have a right to own fundamental biological components of life. This has been a factor influencing sub-Saharan Africa's stand on Article 27(3)b of the WTO TRIPS agreement, which states that there is to be no IP protection for life forms (WTO 1994).

It is estimated that the countries of the Organization for Economic Cooperation and Development hold 97 percent of all patents and global corporations 90 percent of all technology and product patents (11) related to living materials. This lopsided ownership of living materials is a potential source of contention, particularly because of the monopoly it provides to only a few foreign companies.

On the other hand, multinationals do spend enormous resources to develop improved agricultural products. IPR form the core of their financial base and may even catalyze mergers, business deals, and ascription of status. This not withstanding, there is a growing need for partnerships and collaboration between African institutions and these multinationals in the area of technology transfer. IPR are needed to facilitate agreements and ensure an environment of trust. The basic fact is that no company that had spent large sums of money would risk collaboration if protection of its product was uncertain. Research is expensive and may require considerable time input. It requires the use of skills and costly equipment that push up the value of the final product. Compensation for such involvement becomes a necessity, and IPR may serve as a medium for negotiations and reward.

During the October 2002 World Summit on Sustainable Development in Johannesburg, South Africa, heated debates occurred in various forums on the ills of IP as a medium for trade. Examples from group discussions can be found in the brief provided by Genetic Resources International or GRAIN (Kuyek 2002). Claims were made that the multinational seed industry's expansion into Africa had come with intense pressure in favor of IPR, but with no intention to make the technology freely available to farmers. Views expressed at this meeting were that African agriculture does not require IPR because such agriculture is led by farmers, funded by the public sector, and based on collective knowledge. Anti-IPR activists claimed that protection regimes undermine farmers' rights, foster dependence on foreign companies, allow piracy of farmer-developed crops, and threaten food security and agrobiodiversity. But contrasting views were that because of the need to increase productivity, the situation in Africa is no longer static; it is evolving all the time. Local companies, national research institutions, nongovernmental organizations, and farmers' associations are increasingly engaging in biotechnology and other improved agricultural techniques such as tissue culture and marker-assisted selection for higher agricultural yields (Persley and MacIntyre 2001; Persley 1999; Ismael, Benet, and Morse 2001; Bennet 2003; KUZA 2002; Mugabe 2003). Soon genetic modification will become common (University of California–San Diego and Africa Bio 2002).

The Conceptual Framework and Policy Trade-offs

The numerous pros and cons of IP and biotechnology in agriculture clearly underscore the need for comprehensive policy guidelines, not only as a prerequisite for the application of GM technology in food production but also for public assurance of its safety.

The effect of IP on the costs of GM technology is recognized as a potential hindrance to its application in Africa. This concern is shared not only by African authorities but also by international research organizations and some multinational companies (*Genet Archive* 2003; U.S. Embassy, Tokyo, 2003). Apart from straightforward negotiations between potential users and IPR owners, in which the IP may be acquired through contractual licensing, outright purchase, or partnerships, the need to minimize costs, particularly to deserving poverty-stricken developing countries, may require goodwill arrangements including donations. In view of this, a concerted effort appears to be in the making through the recently established African Agricultural Technology Foundation (AATF). Supported by the Rockefeller Foundation and set up in Nairobi, Kenya, under an African-controlled board, the AATF has an ambitious mandate to link the needs of resource-poor farmers with potential technologies acquired through royalty-free licenses, agreements, and contracts. It is expected that multinationals will line up to donate technologies to this noble cause.

Positive reactions to the AATF from corporations such as Monsanto, Dupont, Syngenta, and Dow Agro Sciences demonstrate the goodwill internationally, but it is yet to be seen what impact this approach will have on GM acceptance in Africa and how soon benefits can be felt. Biosafety concerns and lack of biotechnology policies are likely to impede developments.

Several other agencies are involved in the brokerage or application of modern technologies for Africa's agriculture. These include the International Service for the Acquisition of Agro-biotech Applications, the Collaborative Agricultural Biotechnology Initiative of the U.S. Agency for International Development and the Consultative Group on International Agricultural Research (CGIAR). The latter's broad mandate includes mobilization of cutting-edge science to reduce hunger and poverty, improve nutrition and health, and protect the environment. Made up of 16 international agricultural research centers and working in 150 countries, the CGIAR has had a significant impact in some sub-Saharan African countries, where new varieties of cereal and lentil crops are increasingly being grown by farmers. New programs such as those to develop insect-resistant maize, quality protein maize, and *Striga*-resistant and viral-resistant cassava and sweet potatoes are bound to have a positive impact on the economies of small-scale poor farmers.

Ongoing lab tests and research on *Bacillus thuringensis* (*Bt*) maize in Kenya and Zimbabwe under the IRMA (Insect Resistant Maize for Africa) project of the International Maize and Wheat Improvement Center of Mexico (CIMMYT) are forerunners of increased GM activity in sub-Saharan Africa (IRMA 2002). In this case experimentation is being carried out with *Bt* genes found to be active against stem borers, which in Kenya reduce maize production by more than 20 percent. *Bt* genes developed by the CIMMYT, in combination with cry genes from Canada and Centre de Coopération Internationale en Recherche Agronomique pour le Développement (CIRAD), are being evaluated for their effectiveness against African stem borers. In such cases IPR implications have to be addressed.

For example, it is necessary to determine whether the required technology is under protection or whether the protection has expired (as it does after 20 years for patents), in which case it is in the public domain and can be used freely without reference to the owner. Moreover, IPR are territorial, and if a technology is not protected in a particular country by designation, it can be used in that country without reference or remuneration to the owner of the technology. Therefore African countries stand to benefit from the many technologies available globally at minimum cost.

A search of the IP databases at the Kenya Industrial Property Institute and at the Harare-based ARIPO reveals that the cry genes used in the IRMA project are not protected in Kenya. Under the principles of IP protection, such technologies can be used locally without compensation to the owner of the patent. The current mood of multinationals encourages donations or availability of technologies to developing countries at no cost or at low cost.

With such flurry of goodwill among multinationals, international research agencies, and benevolent brokers, it is imperative that African countries be alert and have the correct tools to assess what is good for them. Not every technology for food production is desirable. An example is the use of the infamous "terminator gene," which was the subject of a hue and cry voiced a few years ago (Oliver et al. 1998; Deak 1999; *RAFI Communique* 2000). Both scientists and the public—who may or may not have understood the essence of the problem—objected simply because they smelled something wrong with a technology that would interfere with self-reproduction and the perpetuation of biological material. What is most critical, however, is that African countries have the capacity to decide what is and what is not good technology for them and be able to accurately defend their position. Otherwise, the recipient of a goodwill donation of IP could be the loser in the absence of informed assessment. This again calls for credible biosafety and IP policies to guide the adoption of technology for increased agricultural productivity.

This does not in any way discredit the goodwill of companies and agencies that participate in efforts to address the food crisis in Africa in a benevolent manner. In fact it would be sad to discourage such involvement through careless activism. There is a need for close collaboration among all partners, policymakers included, in the promotion of biotechnologies for food security in Africa for win-win outcomes.

As the previous observations and examples show, it is possible for institutions in Africa to acquire needed agrobiotechnologies cheaply for their food production programs by making use of technologies not protected in their respective countries or those in the public domain. It is also worth noting that for centers under the mandate of the CGIAR system the research performed by the centers is to be used for the alleviation of poverty in resource-poor countries, so any IPR claimed for the centers' products should be free of charge to the system's target countries.

In this respect, the CGIAR centers are bound by the International Treaty on Plant Genetic Resources for Food and Agriculture (ITPGR), which was agreed to by member countries of the Food and Agriculture Organization in 2001. The ITPGR requires that certain genetic materials held by the centers be designated to remain in the public domain for free access by the world community.

It is in this light that networking on biotechnology issues in Africa is absolutely important. Established awareness and public education networks such as the

African Biotechnology Stakeholders forum (ABSF) and Africa Bio have a critical role to play in the sensitization of policymakers, the public, and multinationals on trends in biotechnology that might affect them. To date one can say that these networks have made a formidable first step in this endeavor by delivering information on the initial concepts of biotechnology. The ABSF should be commended for facilitating discussions on biosafety policies in Africa by means of its outreach activities involving parliamentarians, reporters, scientists, and policymakers.

At a different level, nongovernmental organizations in Africa must take up the mantle and get involved at the level of research and transfer of technology, as well as at the advocacy level. Thus activities spearheaded by the Biotechnology Trust of Zimbabwe, the Biotechnology Trust of Africa, A Harvest, the National Biotechnology Development Agency of Nigeria, the Association for Strengthening Agricultural Research in Eastern and Central Africa, and national agricultural research institutes are continuing to provide the farmer-scientist participation that is vital to the better understanding, transfer, and use of biotechnologies.

Research, Capacity Building, and Communication

Due to the importance of IPR as the vehicle for innovation, there is an urgent need for increased capacity in this area within southern African countries. Training in IPR issues takes a long time, especially if one considers the need for skillful agents either to construct patent applications or examine the details of applications for the purpose of registration or for determination of the IP status of a technology. In either case, one has to acquire skills in scientific, legal, and other areas relevant to the administration of IP generally or to awareness creation.

IP offices in southern African countries are scantly staffed. One reason is that governments do not appreciate the importance of such offices. Pressure must now be put on governments to increase the capacities of IP offices in the face of increasing global trade requirements and for national application of IP systems for development. For this to be achieved, governments must allocate adequate funding for staff development and for the effective administration of IP offices.

Governments need to enact or amend various laws to accommodate changes in the local, regional, and international scene, including conformance to the TRIPS agreement. However, it should be understood that IPR should be exercised coherently to the mutual benefit of rights holders and consumers. Regional and international laws on IPR should balance the rights and duties of rights holders vis-à-vis the poor. The laws should reflect the needs of developing countries, particularly their impact on the social and economic development of these countries. In this regard, various international bodies on IPR should work closely with all relevant stakeholders to ensure that the laws do not conflict with public interests.

Outreach activities to give correct information to the public are absolutely necessary. In this respect, there is a need for training of officers and media reporters on issues concerning biotechnology and IPR. A great deal of harm has been done by sensational and inaccurate reporting in southern African countries. Public opinion has been set so negatively that a greater effort is needed to provide objective analysis of biotechnology, IPR, and genetically modified organisms (GMOs). Most important, accurate information and awareness need to be provided to government officials and consumers who have to make decisions as to whether GM technology is needed and is a safe way to enhance food security in southern Africa.

Ethical Issues

Scientific discovery is supported and permeated by moral values. This matters in different ways, depending upon the scientist's social role. At its core, science is an expression of some of our most cherished values. The public largely trusts scientists, and scientists must in turn act as good stewards of this trust. In many African countries a highly disturbing ethical issue related to IPR is raised by the prospect that scientists in industrial countries might patent naturally occurring organisms in developing countries. At issue are access, sharing of benefits, and scope of patents. Is there scope for repatriation of (or compensation for) germ plasm? What are the implications for African countries given their limited capacity to engage with the rest of the world? Is there scope for compensation based on moral pressure? These questions have yet to be consistently posed or answered, but that is likely to change in the very near future.

Conclusions and Recommendations

The preceding analysis suggests the following conclusions and recommendations for southern Africa:

1. Southern African countries have an urgent need for comprehensive policy guidelines for biotechnology application that target biosafety laws and provide clear directions on the handling of GMOs.

2. These countries have an equally great need for policies on IPR that define the role of protection in agricultural inventions, including the desired extent and use of IPR as well as cost and access implications.

3. Attention should be given to capacity development to provide the skills needed for policy development, enactment of laws, and implementation of technologies for increased agricultural production and food self-sufficiency.

4. Partnerships should be encouraged between stakeholders, including multinational companies, international agencies, national research institutions, companies, and nongovernmental organizations, for enhancement of technology transfer to address food security in southern Africa.

5. It is key to create an awareness of the role of biotechnology and its potential impact on food security for southern African countries. Therefore, it will be advantageous to encourage networking and the use of local groups in advocacy and awareness creation efforts aimed at developing an informed society.

6. Southern African governments should ensure the provision of funding for capacity building and the development of laws, policies, structures, and an environment altogether conducive to increased food production.

References

Bennet, A. 2003. *The impact of Bt cotton on small holder production in the Makathini Flats, South Africa.* Monsanto SA (Pty) Ltd.

CGIAR (Consultative Group on International Agricultural Research). 2002. *Challenge programmes. Earth Summit 2002 and the future harvest centres.* Washington, DC: CGIAR Secretariat. August.

Deak, F., ed. 1999. Monsato to finalise terminator control. May 17. thinker@universe.com.

Ekpere, J. A. 2000. *The African model law.* Organisation of African unity. Addis Ababa, Ethiopia.

Friends of the Earth International. 2003. *Agricultural biotechnology: GMO contamination around the world.* 2nd ed. Amsterdam.

Genet Archive. 2003. To feed hungry Africans, 7 business/biotech companies set up foundation for seed technology transfer. March 12.

Hivos and Friends of the Earth International. 2003. *The world as a testing ground: Risks of genetic engineering in agriculture.* London: Friends of the Environment.

IRMA (Insect Resistant Maize for Africa). 2002. *Annual report.* IRMA project document no. 10. Mexico City.

Ismael, Y., R. Benet, and S. Morse. 2001. Arm level impact of Bt in South Africa. *Biotechnology and Development Monitor* 48 (December).

Johns Hopkins University School of Public Health, Center for Communication Programs. 2000. *Population reports,* series M, no. 15, vol. 23, no. 3 (fall).

Kabare, J. N., and J. M. Wekundah. 2002. *Proceedings of a regional workshop on biotechnology and IPR.* Biotechnology Trust Africa, Barcelona. March.

Kuyek, D. 2002. Intellectual property rights in African agriculture: Implications for small scale farmers. Genetic Resources International (GRAIN). August.

KUZA. 2002. Advancing agriculture in Africa. *KUZA Newsletter,* no. 14.

Lauderdale, J. 2000. *CIMMYT improves nutritional quality of maize.* Mexico City: CIMMYT (International Maize and Wheat Improvement Center).

Mugabe, J. 2003. *Keeping hunger at bay: Genetic engineering and food security in sub-Saharan Africa.* Nairobi, Kenya: African Technology Policy Studies Network.

Oliver, M., J. E. Quisenberry, N. L. Glover Trolinder, and D. L. Keim. 1998. U.S. Patent: Control of plant gene expression. Patent no. 5,723,765. March 3.

Persley, G. 1999. *Letter to a minister.* 2020 Vision Brief no. 10. Washington, DC: International Food Policy Research Institute.

Persley, G. J., and L. R. MacIntyre, eds. 2001. Agricultural biotechnology country case studies: A decade of development. Wallingford, CT, USA: CAB International.

RAFI Communique. 2000. The terminator gene is still on fast track. February–March.

University of California–San Diego, Center for Molecular Agriculture and Africa Bio. 2002. Food from genetically improved crops in Africa. April.

U.S. Embassy, Tokyo. 2003. USAID announces international biotech collaboration. October 4.

WTO (World Trade Organization). 1994. *Agreement on trade-related aspects of intellectual property rights.* Annex 1C of the Marrakech Agreement establishing the World Trade Organization, signed in Marrakech, Morocco, on April 15, 1994. Geneva.

Chapter 7

Trade Policy

Moono Mupotola

The 2002 food crisis in southern Africa, which was exacerbated by the reluctance of the countries to accept genetically modified (GM) maize food aid, highlighted the need for the region to address the trade-related issues raised by biotechnology, especially given the move toward formation of a free trade area (FTA) by 2008. One feature of the FTA will be the free movement of agricultural products across borders.

This chapter attempts to highlight the key issues related to biotechnology and trade, particularly as they relate to the agricultural sector of the Southern African Development Community (SADC) region. The questions asked include whether there are opportunities for the SADC given that trade in agricultural commodities plays an important role in the economies of these countries. Furthermore, attempts are made to address some of the concerns surrounding biotechnology and biosafety in the SADC.

GMOs and International Trade in Agricultural Products

Trade in genetically modified organisms (GMOs) is highly influenced by the international regulations that govern world trade. The major agricultural countries are countries in the north that have had a tremendous influence in shaping the nature of the world trading system. With the introduction of GM products, it has become apparent that the scope exists for developing countries to benefit from this technology through higher yields, lower production costs resulting from reduction in pesticide use, and, in the case of net food-importing countries, the ability to source cheap food.

World Production of and Trade in GMO Crops

Production of biotechnology crops is concentrated in a few countries, of which developing countries account for 15 percent of the area planted with transgenic varieties. The United States is by far the largest, accounting for at least 68 percent of production, followed by Argentina (23 percent), Canada (7 percent), and China (1 percent). Other countries therefore produce just 1 percent of the total output. The greatest area is devoted to soybeans, cotton, corn, and rapeseed—that is, commodities that are also traded internationally. As shown in Table 7.1, the production of biotechnology crops is concentrated in a few countries; however, the number of importing countries is large in comparison. This illustrates that there is a large market for these commodities given that some GM commodities are processed and their extractions, such as edible oils, cornmeal, and soybean proteins, are used as ingredients in more than 70 percent of the processed foods available in most developed-country markets (Phillips 2003). The International Seed Federation estimated that the value of world trade in GM seed was US$4.5 billion in 2004 (Oxfam 1999).

World trade in GM commodities is concentrated in soybean products (Table 7.2). This is not surprising, because soybeans account for 58 percent of the area planted in GM crops worldwide, followed by corn (23 percent), cotton (12 percent), and canola (6 percent) (Diaz-Bonilla 2002).

Table 7.1 Production of and trade in genetically modified agricultural food products, 2000

Crop	Number of producing countries	Percent of exports from GM producer	Number of importing countries
Maize or corn	8	85	168
Soybeans	6	88	114
Canola	2	50	68

Source: Phillips 2003.

Table 7.2 Estimated percentage of international trade in genetically modified organisms, 2000

Product	Percent
Cottonseed cake	10–20
Cottonseed oil	15–25
Corn	10–20
Soybean cake	25–35
Soybean oil	25–35

Source: Diaz-Bonilla 2002.

As Phillips (2003) observes, those countries adopting biotechnology methods tend to be traditional exporters, and they "thereby increase their exportable surplus, depressing world prices and making nonadopting importing producers less competitive." This is indeed a worrying trend for African countries that want to compete internationally in an already "price distorted" international trading system in which world prices are depressed because developed countries still have highly protected markets and subsidize their farmers.

The International Legal Framework
One of the cornerstones of the Marrakesh Agreement and the subsequent establishment of the World Trade Organisation (WTO) was the introduction of trade regulations for agricultural products. When they become members of the WTO, countries are obliged to follow the rules that are set out in the various agreements that pertain to trade in agricultural products.

There are three legal frameworks relevant to trade in GMO products under the WTO. The first is the Sanitary and Phytosanitary Measures Agreement (SPS), which specifically relates to food safety, as well as plant and animal health. The second is the Agreement on Technical Barriers to Trade (TBT), which deals with technical regulations, voluntary standards, and compliance procedures except when these are defined as SPS measures (Anderson and Nielsen 2000). The third is the Agreement on Trade Related Aspects of Intellectual Property Rights (TRIPS), which sets out standards for intellectual property rights (IPRs) that members must follow.

International standards are encouraged in both agreements where they exist, although the SPS agreement permits the use of risk assessments where international standards do not exist. The TBT agreement is more flexible, as it allows member countries to decide against an international standard based on its own unique situation, such as national security interests. While the SPS agreement allows for risk assessments in the absence of an international standard, it emphasizes that such assessments must be based on science and should not be used as barriers to trade. Yet the major area of contention regarding GMOs is precisely the lack of an international standard, which gives member countries room to adopt trade-restrictive measures regarding trade in GMO products.

The TRIPS agreement (WTO 1994), particularly Article 27(1), specifies that member countries must patent any invention, "whether products or processes in all fields of technology," and that these must be transparent and for a period of 20 years from the filing date. The TRIPS agreement also enables the patent holder to exclude others from making, using, or selling the invention. However, a major weakness is that the agreement does not define an invention.

Complicating matters further is Convention on Biodiversity and its Cartegena Protocol on Biosafety (CPB), which is in conflict with the WTO agreements. The CPB provides for the "safe transfer, handling and use of GMOs that may have adverse effects on the conservation and the sustainable use of biological diversity, taking also into account risks to human health, and specifically focusing on transboundary movements" (Diaz-Bonilla 2002). The major scope of the CPB is the "precautionary principle," its relationship with other agreements, and liability. Although the CPB is yet to be ratified, the precautionary principle gives discretion to countries to establish standards even without full scientific certainty about the problem concerned and allows countries to decide under what conditions they will accept GM products for domestic release.

GM Controversies and Trade

In 1999 a four-year ban was pronounced on new GM crops in the European Union. This decision has led to strong disagreements between the European Union and the United States over the European Union's regulation of GM foods.[1] The United States claims that these regulations violate free trade agreements; the European Union's counter-position is that free trade is not truly free without informed consent. This position has been further cemented by widespread concern within the European Union about GMOs in terms of environmental protection (in particular, biodiversity) and the health and safety of consumers. Many European consumers are demanding the right to make an informed choice. New EU regulations should require strict labeling and traceability of all foods and animal feed containing more than 0.5 percent GM ingredients. EU directives, such as Directive 2001/18/EC, were designed to require authorization for placing GMOs on the market, in accordance with the European Union's precautionary principle.

At the end of 2002, EU environment ministers agreed to new controls on GMOs that could eventually lead the 15-member bloc to reopen its markets to GM foods. The EU ministers agreed to new labeling controls for GM goods, which will have to carry a special harmless DNA sequence (a DNA bar code) identifying the origin of the crops; making it easier for regulators to spot contaminated crops, feed, or food; and enabling products to be withdrawn from the food chain should problems arise. A series of additional sequences of DNA with encrypted information about the company or what was done to the product could also be added to provide more data.

Many European consumers are asking for food regulation (demanding labels that identify which foods have been genetically modified), while the American agricultural industry is arguing for free trade (and is strongly opposed to labeling, saying it gives the foods a negative connotation). They claim mandatory labeling could imply that there is something wrong with GM foods, which would also be a trade barrier. Current U.S. laws do not require GM crops to be labeled or traced, because U.S. regulators do not believe that GM crops pose any unique risks compared to conventional foods. Europe answers that the labeling and traceability requirements are not limited to GM food, but will also apply to any agricultural goods. The Americans insisit that what the EU is doing is a breach of WTO rules and is "immoral" because it could lead to starvation in the developing world, as seen in some famine-threatened African countries (e.g., Mozambique, Zambia, and Zimbabwe) that refuse to accept U.S. food aid because it includes GM food.

In May 2003 the George W. Bush administration officially accused the European Union of violating international trade agreements by blocking imports of U.S. farm products through its long-standing ban on GM food. A formal complaint challenging the moratorium was filed with the WTO after months of negotiations trying to get it lifted voluntarily. The complaint was also filed by Argentina, Australia, Canada, Chile, Colombia, Egypt, El Salvador, Honduras, Mexico, New Zealand, Peru, and Uruguay. The formal WTO case challenging the EU regulatory system was in particular supported by U.S. biotechnology giants such as Monsanto and Aventis and by big agricultural groups such as the National Corn Growers Association.

In June 2003 the European Union Parliament ratified a three-year-old UN biosafety protocol regulating international trade in genetically modified food, which was expected to come into force in the fall of 2003 because the necessary number of ratifiers was reached in May 2003. The protocol lets countries ban imports of a GM product if they feel there is not enough scientific evidence that the product is safe, and it requires exporters to label shipments containing genetically altered commodities such as corn or cotton. It makes clear that products from new technologies must be based on the precautionary principle and allow developing nations to balance public health against economic benefits.

On July 2, 2003, the European Union Parliament approved two laws that will allow the European Union to lift its controversial ban on GM food. The first law will require labeling for foods with more than 0.9 percent GMO content. It will be applied to human food and animal feed as well. However, animal feed containing transgenic cereals will not be included in the labeling. The second law will make mandatory the labeling of any food contaminated by GMOs not authorized (in the European Union) if the amount is more than 0.5 percent of the total. This amount will be set for three years. After three years, all food contaminated with nonauthorized GMOs will be banned. Traceability of GMO products will be mandatory from sowing to final product. At the time that the ban was imposed, it was expected to be lifted in the fall of 2003.

In May 2004, the European Union lifted the ban on GM food imports by endorsing an application by a Swiss biotechnology company, Syngenta, to import GM corn. The future remains unclear, however. The ban was lifted despite intense public opinion against such an action.

The SADC: Agricultural Production and Trade

Production

Agriculture remains a dominant economic activity in southern Africa. However, partly because most of the region's major staples—such as maize, cassava, and millet—are grown in subsistence-oriented systems, recent droughts and floods have depressed output significantly and threatened food security in a number of countries simultaneously.

According to the SADC Regional Early Warning Unit (REWU), the regional cereal production for the 2001/02 season of 21.75 million tonnes was below the five-year average of 22.44 million tonnes. A further 1.2 million metric tonnes of food was needed in the six countries most affected. In addition, it should be emphasized that a key characteristic of agriculture in the SADC is low productivity. A key question, therefore, is whether the SADC countries can raise productivity to avert dependence on cereal imports from outside the region and increase trade, particularly given the abundance of arable land in some countries.

Trade

The SADC's major trading partner is the European Union, while the United States, Japan, and the Far East are also important markets. Except for South Africa, which has an FTA with the European Union, most of the SADC countries have benefited under the European Union's preferential trade agreement with the countries of the African, Caribbean, and Pacific region known as the Cotonou Agreement.

Under the beef and veal protocol of the Cotonou Agreement, four countries of the SADC, namely Botswana, Namibia, Swaziland, and Zimbabwe, can export a specified tonnage of beef into the lucrative EU market paying only 8 percent duty. The sugar protocol also gives several countries in the SADC, namely Malawi, Mauritius, Swaziland, Zambia, and Zimbabwe, preferential market access to the EU market. There are other provisions as well, such as that granting preferential market access for grapes, of which Namibia is the main beneficiary. Because preferences are restricted to a specific period of the year, Namibia is the only country in the southern hemisphere that has access to this market at a time that coincides with its harvesting season.

Table 7.3 Fast-growing agricultural product areas under the African Growth and Opportunity Act

Product	Percentage increase, 2001–01
Cut flowers	2,258
Frozen vegetables	689
Dates, figs, pineapples	1,468
Fruit juices	1,342

Source: United States Trade Commission, Washington, D.C.

More recently the African Growth and Opportunity Act has provided the SADC countries (except Zimbabwe) preferential market access. Although textiles dominate, trade in other agricultural products is growing. As Table 7.3 illustrates, the fastest-growing agricultural exports are high-value products such as cut flowers, dates, figs, pineapples, and fruit juices.

In September 2000 the SADC launched a trade protocol that aims to establish an FTA by the year 2008. During the past decade intra-SADC trade grew faster than did total SADC trade. It is estimated that between 1991 and 1996 total SADC trade grew at a rate of 13.8 percent, while intra-SADC exports and imports grew at a rate of 23.1 percent and imports at a rate of 17.7 percent (SADC-UNDP 2000). Although this shows some degree of integration, countries of the Southern African Customs Union (SACU)—Botswana, Lesotho, Namibia, South Africa, and Swaziland—dominate these trade flows. South Africa dominates, contributing 94 percent of all SACU exports and 98 percent of total SACU imports (SADC REWU 2002). In 1997 the five SACU countries accounted for 41 percent of SADC exports and 48 percent of SADC imports.

The SADC's major agricultural exports are cash crops such as tea, coffee, tobacco, sugar, horticultural products, cotton, maize, livestock, and livestock products. Imports comprise mainly cereals such as maize, agricultural inputs, and a range of food commodities.

Production of GM Crops in the SADC

South Africa is the only country in the SADC that grows GM crops at a commercial level. Three crops, cotton, maize, and soybeans, which may be insect- or herbicide-resistant, have been approved for commercial release. Currently approximately 200,000 hectares of GM crops are grown in South Africa in areas such as the northern provinces, KwaZulu/Natal, and the Orange Free State. An estimated 28 percent of the cotton planted in South Africa is GM, while GM white maize varieties are about 6 percent of the total maize grown.

Cotton

Bacillus thuringiensis (*Bt*) cotton is grown on 100,000 hectares by 1,530 commercial farmers and 3,000 small-scale farmers (Kuyek 2002). The production of *Bt* cotton is often hailed as a success story, particularly for small-scale farmers. In fact it is estimated that 7 of every 10 South African farmers have switched to GM varieties (Hetherington 2003). Some of the positives listed by South African farmers are that by using *Bt* cotton they have decreased their production costs due to less use of pesticides and that the zero tillage required allows for greater water retention in the soil (Hetherington 2003). Yet the success of small-scale farmers, particularly in South Africa, and the fact that the country imports about 50 percent of its cotton to meet its requirements could be an incentive for other SADC countries, particularly given the advantages provided by the trade protocol in terms of market access.

Maize

Most GM varieties in the SADC have focused on reducing pesticide usage. Trials in South Africa show that the yield advantage of using GM varieties is quite small and varies between a decrease of 7 percent and an increase of 13 percent (MAWRD 2002). In 1999 South Africa planted 50,000 hectares of *Bt* maize. One criticism of *Bt* maize in South Africa is that it has been developed only for commercial farmers and not for small-scale farmers (Kuyek 2002).

Generally the limited research into maize varieties used by small-scale farmers is not limited to South Africa alone. The International Center for Maize and Wheat, with support from the Novartis Foundation, is working to develop *Bt* maize varieties for small-scale farmers in Africa. However, there is a potential problem related to intellectual property rights, as Novartis donated its *Bt* technology for research purposes only.

Policy Issues and Trade-offs

Two sets of policy issues and trade-offs emerge; one set relates to imports, the other to exports. With regard to imports, the key questions are these: How can countries take advantage of cheap GM grain while guarding against possible human health effects? Which are the major traded commodities for which GMOs are important? Are these crops potential import crops for southern African countries?

With respect to exports, it is clear that biotechnology and GMOs may increase productivity and make commodities more price-competitive on world markets. But this may come at the cost of a higher risk of reduced access to key markets, especially in Europe, where consumer sentiment against GMOs is likely to remain high well into the future. Again, which are the major traded commodities for which GMOs are important? Are these crops potential export crops for southern African countries?

As noted earlier, the SADC's major agricultural imports are cereals such as maize, agricultural inputs, and a range of food commodities. Her exports are cash crops such as tea, coffee, tobacco, sugar, horticultural products, cotton, maize, livestock, and livestock products.

The recent food crisis in the SADC region has highlighted that food security is still a major problem in the region. As the SADC moves toward deeper integration through trade, whether the production of GM crops could alleviate the food security situation in the region is an open question. There are certainly advantages to GM technologies. Even skeptical organizations such as Oxfam agree that "GM technology offers potential to contribute to higher yields and crop productivity of interest to poor farmers and that these opportunities should be researched" (Oxfam 1999). The SADC has among its members net food-importing countries such as Botswana, Lesotho, and Namibia. Because of their climatic conditions, these countries are unlikely ever to be self-sufficient in food production. Cheaper food imports are to their advantage. Moreover, some GM products are extensively processed and are used as ingredients for other products. Some SADC members may question the safety of consuming GM maize; however, there are ways of mitigating the introduction of GM varieties into the environment. The five countries that accepted GM maize meal during the recent food crisis agreed to have the grain milled at specific points before it was distributed nationwide. But it is clear that the information needed to resolve the import-related policy trade-off noted earlier is still unavailable.

The case of *Bt* cotton allows some preliminary positive responses to the question of exports. Clear benefits appear to be accruing to a wide spectrum of farmers, including smallholders, due to increased yield and lower production costs. Byproducts such as cottonseed cake and cottonseed oil also present further income-generating opportunities. But the impact *Bt* cotton may have on the environment remains unclear.

The case of beef is rather different. To protect lucrative markets, farmers exporting beef to the European Union have ventured into traceability programs—for instance, FanMeat in Namibia—to satisfy the consumer demands of that market. While the European Union maintains that its does not prohibit the use of GM feed for cattle, some SADC countries maintain that some European buyers insist on certification that GM feed was not used. Therefore most countries would rather take precautionary measures instead. Moreover, the European Union has introduced a labeling law that requires commodities with a GM content of as little 1 percent to be labeled. While the U.S. market may be an option, phytosanitary regulations are a hindrance, as risk assessments must be conducted, and these may take several years and are very expensive.

Other markets for beef can be sought; however, the lack of uniformity in sanitary measures in developed-country markets can hinder diversification into other markets. An important aspect is that the EU market offers a premium price for cuts exported under the Beef and Veal Protocol within the Cotonou Agreement.

Many SADC countries that have diversified agricultural production have ventured into horticultural products, supplying supermarkets such as Sainsbury's and Tescos in the United Kingdom. These are obvious niche markets that many producers in the SADC would not want to jeopardize. There is also a growing trend in the SADC for exporters to access the organic market, which attracts premium prices. In Zambia, for instance, the Organization of Organic Producers and Processors of Zambia, which has a membership of 100 farmers, exports vegetables, herbs, and coffee to the European Union and the United States.

Soybeans are another potential export crop for the SADC, as soybeans are one of two crops (the other is bananas) that account for 64 percent of developing country crop exports to developed-country markets. The key exporters of GM soybeans are developing countries, notably Argentina and Chile. Within the SADC, South Africa is a key market, as it has a well-developed agribusiness sector.

The major trade-off for countries that embrace biotechnology is therefore the extent to which this may affect trade with the European Union. It is worth pointing out that despite South Africa's relatively long history of producing GM crops, the European Union remains its main trading partner. The key recognition is that South Africa's agricultural production base is diversified and modern.

Conclusions

There are advantages to the use of biotechnology; however, it is not a panacea for alleviating the food security needs in the SADC region. Apart from developing capacities at national levels, the SADC governments should embrace the need to fully participate in the negotiation of various legal instruments that govern international trade in agricultural products.

It is of no use to increase productivity leading to an exportable surplus if a country has no market access. The current trading system is stacked against developing countries. Developed-country markets are highly protected, their farmers are subsidized, they have highly bureaucratic procedures, and they are expensive to access. Countries that have managed to access the EU or U.S. markets have had to spend considerable amounts of money to do so. Therefore, without addressing issues such as export subsidies and their devastating effects on world prices, trading with these developing countries would not be of considerable benefit to developing countries. It is in the SADC's interest that member countries act as a cohesive group in areas of mutual interest during negotiations of international agreements. If they could influence the overall world trading system, the SADC countries would not have to rely on preferential market access opportunities alone.

Note

1. The following discussion of GM-related controversies in trade is drawn from a range of sources available on the Internet. An important set of sources can be found at the following Web site: http://www.fact-index.com/t/tr/trade_war_over_genetically_modified_food.html.

References

Anderson, K., and C. Nielsen. 2000. *GMOs, food safety and the environment: What role for trade policy and the WTO?* Policy Discussion Paper no. 0034. Centre for International Economic Studies, Adelaide University, Australia.

Diaz-Bonilla, E. 2002. Biotechnology and international trade. Power Point presentation at a conference on Agricultural Biotechnology: Can it help reduce hunger in Africa? Washington, DC, March 5–7.

Hetherington, A. 2003. GM pays the bills. *Mail and Guardian* (South Africa), March 20–27.

Kuyek, D. 2002. *Genetically modified crops in Africa: Implications for small farmers.* Genetic Resources International (GRAIN), Barcelona, Spain.

MAWRD (Ministry of Agriculture, Water and Rural Development, Namibia). 2002. *A cost benefit analysis of the utilization of GMOs in the production of Namibia agricultural products for local and international consumption.* Draft final report. Namibia Resource Consultants, Windhoek. October.

Oxfam. 1999. *Genetically modified crops, world trade and food security.* Position paper. www.oxfam.org.uk.

Phillips, P. 2003. *Policy, national regulation and international standards for GM foods.* International Food Policy Research Institute, Research at a Glance, Briefs 1–6, Washington, DC.

SADC REWU (Regional Early Warning Unit). 2002. *SADC Food Security Quarterly Bulletin, Zimbabwe.* October 31.

SADC-UNDP (UN Development Program). 2000. *SADC regional human development report 2000: Challenges and opportunities for regional integration.* Harare, Zimbabwe: SAPES Books.

WTO (World Trade Organization). 1994. *Agreement on trade-related aspects of intellectual property rights.* Annex 1C of the Marrakech Agreement establishing the World Trade Organization, signed in Marrakech, Morocco, on April 15, 1994. Geneva.

Chapter 8

Lessons and Recommendations

Klaus von Grebmer and Steven Were Omamo

Biotechnology, like a host of other complex and multidimensional issues in the development field, has been characterized by marked conflict between different ethical and ideological perspectives. What has contributed to making the differences so entrenched are the profound uncertainties regarding who will benefit and who may lose from the technology, what its unforeseen consequences may be, how long it will take for the impacts to be discovered, whether the effects can be known before irreparable harm is done, and who will make the decisions. With these questions remaining by and large unanswered, different deep-seated beliefs about technology, nature, the global order, and the meaning of development on the part of the various stakeholders have come into play, increasing the intensity of the dispute and making it seem irreconcilable at times.

In today's globalizing economy, a country, particularly a developing one, will not be able to survive unless it adopts or accommodates to genetic engineering in agriculture. If it is to compete internationally, it will have to adopt biotechnology for production. For many countries, not investing in biotechnology may also mean greater environmental degradation and food insecurity. It can no longer even be considered an option, because developing-country institutions have been conducting research on the technology for almost two decades in some cases and have developed products that are already fundamentally transforming agricultural production, trade, and consumption. At the very least, a country will face difficulties in seeking to keep genetically modified (GM) crops and foods out of its borders as international economic agreements and world trends pressure it to accept them.

Biotechnology has the potential to be a key driver of development, poverty alleviation, food security, and natural resource conservation in the developing world if practiced responsibly. And while questions remain about for whom and for what biotechnology will ultimately be employed, more immediate and pressing ones exist, the answers to which in fact must be pursued in a concerted and collaborative manner if we are to ensure that the technology benefits and does not harm society and the earth. Some of the questions are these: What biosafety regulatory frameworks should be established? What policies are required to guarantee that the production of GM crops serves poor farmers and consumers? And what research and information are needed to develop frameworks and policies on these issues and other important ones?

The primary motivation for the 2003 Regional Policy Dialogue on Biotechnology, Agriculture, and Food Security in Southern Africa was the food crisis facing the region. Historically weak policies to encourage and enable increased agricultural production among smallholders, coupled with environmental shocks, had brought a severe shortage in food crops and left millions of people at risk of starvation. The crisis, which was only slightly alleviated owing to the inadequate responses on the part of the governments in the region, underscored for many in the development community the need for wider agricultural biotechnology adoption and dissemination in southern Africa. The conflict over the GM food aid that arose as these governments, donor countries, and international organizations attempted to address the situation revealed that, regardless of whether the aid was accepted in this case, it was imperative for the countries of the region, and indeed for all developing countries, to have a biosafety system to scientifically evaluate the risks of GM products for their respective national contexts.

Yet today the region as a whole is not far along the road of biotechnology development and assessment. Modern biotechnological techniques are being employed in only a few southern African countries, namely, Malawi, South Africa, Zimbabwe, and to a lesser extent Mauritius and Zambia. Of these countries, only South Africa has reached the commercialization stage for genetically engineered (GE) goods. The others have either only recently approved contained crop trials or do not yet have the regulatory or scientific capacity necessary to conduct such trials.

The food crisis in the region fundamentally and irreversibly altered the content and nature of the debate on how to respond to such crises. But biotechnology has also changed the debate on how long-term agricultural growth and food security can be achieved with technological advances in agriculture. To many stakeholders both in the region and outside it, GM food aid signaled the likelihood of the production of GM crops in the region not far in the future. Generally, while some welcome this prospect, others see this potential development as adverse. Both groups, however, are concerned about the numerous uncertainties regarding the relevance, efficacy, sustainability, and safety of the technologies.

This chapter seeks to draw from the preceding chapters some of their lessons and recommendations for the future for consideration by stakeholders in southern Africa and the wider agricultural development community that needs to support them. To properly address the uncertainties that biotechnology raises, generate information, and ensure that the technology serves the needs of the poor in southern Africa in an environmentally sustainable way, the multistakeholder dialogue begun in earnest at the Regional Dialogue held in Johannesburg—a dialogue at the national and regional levels involving public- and private-sector bodies and nongovernmental organizations (NGOs) concerned about the issues the technology raises—will have to be expanded and sustained. Through involving groups from civil society this dialogue might attain characteristics of being a societywide process. The conflicts over biotechnology both at the global level and in southern Africa are deep, and without a consensus-building process it is unlikely that biotechnology will move in any direction. The decisions each country and the region as a whole will ultimately make on the issues is another question. But what are urgently required in the debate at this point are greater awareness, information, and understanding, which research can further, as well as more clarity on the measures that can be adopted on the more practical issues, many of which need to be implemented immediately. These include measures related to biosafety, trade-related issues, and biotechnology adoption in the region's agriculture. How to develop capacity for biotechnology governance will be another question the dialogue will be able to inform.

An ongoing regional dialogue will certainly face challenges, because the uncertainties and controversies surrounding the role of biotechnology in agricultural development and food security enhancement are not peculiar to southern Africa, but rather reflect those of the entire global community, and because the need to resolve urgent matters, such as those surrounding biosafety, may work against the process of reaching consensus. However, if the dialogue can serve as a framework for more effectively addressing these matters, and in turn be enriched by the information generated from actions taken, it can sustain the interest and commitment of the stakeholders and more likely direct biotechnology toward reducing hunger and poverty in the region.

Expanding and Sustaining Multistakeholder Processes in the Region

Why are multistakeholder dialogues on biotechnology so important? As a number of the chapters in this book have illustrated, while on the surface the clashes over agricultural biotechnology may appear to be only about the level of protection given the environment or about the procedures and regulations countries must follow, they are fundamentally about differences between disciplinary perspectives, ethical worldviews, and paradigms. Moving toward consensus on the issues will require exploring and finding some common ground between these deeper and more powerful notions, which in large part form the identities of those who hold them.

Differences among informed stakeholders stem to a degree from contrasting disciplinary approaches and methodologies for knowledge generation. Whereas in the biophysical sciences a tight, narrow, and experiment-based hypothesis-testing approach is employed, the social sciences are interested in looser and broader hypotheses on collective behavior for which neither theory nor data provide clear answers on causal relationships. At a more profound level, the reductionism that drives model building and hypothesis testing in the sciences, including the work of some social scientists, is opposed by the more humanities-oriented approaches to social study, in which explanation tends to be built on narrative and ideological perspectives often explicitly inform analysis. In some cases, as in that of environmental advocacy groups, political perspectives and scientific hypothesis-testing approaches merge.

Among these stakeholders and those whom agricultural biotechnology will more directly affect, various competing moral frameworks and cosmologies provide what might be seen as differences in shade. In Chapter 3 Julian Kinderlerer and Mike Adcock point out that in the minds of many people the current food crisis requires that biotechnology be introduced immediately to alleviate the suffering of the hungry. The Nuffield Council on Bioethics argues that developed countries face a compelling moral imperative to make GM crops readily and economically available to developing countries (Nuffield Council 1999). Others might support the use of the technology, but argue that governments and the scientific community have a duty to ensure that it is made available in a responsible way. Still others, distrustful of the technology, believe it is society's obligation to introduce the technology only once the appropriate legislation and regulatory frameworks are in place and risk-benefit assessments have been carried out. For this group developing and using genetically modified organisms (GMOs) are equivalent to "playing God": unnatural acts that can lead to unforeseen negative consequences for humans and the environment and should not be engaged in. It is not only environmental advocacy groups that hold this view: many societies have a deep-rooted belief that tinkering with nature is unacceptable. This view is likely to be as strong in southern African societies as it is in Europe. At stake are different paradigms of human progress and the role of science and technology in human development. In the words of the Nuffield Council, "Proponents of the technology citing practical benefits may have an intrinsic value system that views science and progress as good things in themselves, and opponents may be analysing risks from a world-view that questions the rightness of technological progress."

Principles of justice are involved in this ethical worldview, which seeks answers to questions such as these: Is this new technology likely to increase the gap between the rich and the poor, both within developing countries and between these countries and the developed ones? Will the technology serve those who really need it, the poor? If the technology does enable more efficient and greater food production, will it do so at the expense of those who farm traditionally? Is this acceptable? Should consumers in the developed world eat GM foods if unjust economic and social processes have produced them? The ethical questions are not just about playing God, but about who benefits, by how much, and at what costs. A particular, complex, and normative understanding of the world is at work as each stakeholder deals with the issue of agricultural biotechnology.

Indeed it is not difficult to comprehend why the reactions have been so strong on all sides and why stakeholders inject their positions with their fundamental values. As David Pelletier shows in Chapter 4, although GM proponents in the U.S. government and some outside it claim to be using "sound science," the evidence reveals that the conclusions on the safety of GM crops have been backed up more by appeals to institutional authority than by adherence to the principles of scientific investigation. Pelletier's findings are important and troubling, and have wide and major implications. However, after calmer consideration one might say they are not entirely surprising. Faith in pronouncements claimed to be scientific has declined not only among the formally educated, but among the informally educated as well. Academia is more aware, and even inescapably aware, that ideologies underlie even the most "objective" scholarship, while in the real world people have experienced disillusionment with their leaders that has made them question the truth of official statements. Given the uncertainties involved in biotechnology, the fears to which they give rise, and the principles and rights that are at stake, it is understandable why the conflicts over it have been so great.

When a deeper appreciation of the controversy has been achieved, it becomes imperative that these underlying values, ideologies, and paradigms be addressed if some consensus on the use of biotechnology is to be reached. Furthermore, the intensity of the debate suggests that the key ethical and moral issues ought to be resolved to some extent before agricultural biotechnology is implemented. A multi-stakeholder dialogue therefore needs to include these issues in its agenda in order to bring some resolution to them and to find and maintain a dynamic balance between ethical and technical priorities. There has in fact been a growing recognition of the need, when dealing with scientific questions, to incorporate into the deliberative process broader considerations based on normative concepts. Insights from both the positivist and the normative traditions are becoming increasingly integrated as agencies, stakeholders, and communities seek to develop more productive and appropriate methods for managing the risks and benefits of new technologies.

In Chapter 2 David Matz and Michele Ferenz outline the key conceptual issues in multistakeholder processes and offer various examples of the forms such processes can take. The case studies and the discussion they provide help build an understanding of the kinds of conceptual and practical questions that must be answered to facilitate an effective process. Unfortunately, as Matz and Ferenz state, the various attempts to build consensus on biotechnology in developing countries have not been explicitly conceived or implemented as multistakeholder processes in that they have not been fully cognizant of the central challenges facing such processes. Multistakeholder dialogues are based on the notion that the parties in negotiation almost always have both competing and complementary or compatible interests. The challenge is to structure the negotiations so that these common interests are allowed to emerge and serve as the basis for a mutually beneficial resolution. In short, the negotiation becomes a joint discovery and problem-solving exercise. The key is to focus the discussions on the needs and interests of the stakeholders and the reasons underlying their positions.

From the contributions in this book it is clear that there are essentially four challenges that must be met by a multistakeholder dialogue in southern Africa or by any such process:

- Ensuring that all the relevant parties are involved in negotiations

- Getting accurate scientific and technical information on the table

- Promoting links with official decisionmaking bodies

- Establishing fairness and efficiency as criteria for evaluation of multistakeholder processes

It is clear to those who deal closely with issues related to biotechnology in southern Africa that the debate there is still confined to a very small and select group of stakeholders. In order to ensure a more genuine dialogue at the national or the regional level in southern Africa, organizations representing farmers and the rural poor, including women and consumers, will have to be brought into them. The negotiation process must be accessible to all interested groups and also transparent. Yet while organizations in civil society can provide creative thinking and generate innovative policy options, it will be necessary to verify that they have the requisite capacity to participate actively in the deliberations. The uneven participation of stakeholders is a common problem in such dialogues, and capacity constraints are one of the major obstacles to effective participation. This is a particular problem when stakeholders with vastly different levels of resource endowment come to the table together. The voices and recommendations of members of community-based organizations and NGOs ought to be taken seriously, but for this to be possible they must be well prepared, well organized, and able to remain in the dialogue over a long period of time.

Providing more information for all the participants is also crucial, as is discussed further later. The information must be in a form that all the parties can comprehend. The outcomes of a multistakeholder dialogue are typically not legally binding unless taken up by the relevant governmental authorities. Such a process in southern Africa will complement, not supplant, the established decisionmaking channels. But in order for the dialogue to translate the greater understanding of the issues it achieves into improved policies it will be critical for it to engage and assist those responsible for making decisions on the issues. Finally, monitoring and evaluating technologies and the regulations designed for them will have to be an essential part of any dialogue. However, it will also be vital to monitor and evaluate the dialogue itself, through engaging the participants, in terms of whether it is giving each stakeholder an equal voice, does not have a vested interest behind it, and is actually producing results.

In fact, the aim of a dialogue in the region should not be so much to develop consensus. Rather it should be to agree on the nature of the process that the countries and the region as a whole need to adopt to move toward consensus. What types of processes can be employed? Stakeholders could reflect on the types of dialogues that have been used effectively in other settings and those on biotechnology that are emerging in the region. Developing consensus on the issues will not be an easy task. If the focus is on ensuring a good process instead, positive outcomes will be generated along the way, which in turn will provide stakeholders with an incentive to continue participating in the dialogue. To agree on a process, stakeholders will more specifically need to do the following:

- Resolve to have a learning experience

- Bring those who are not involved in the dialogue to the process (particularly farmers, consumer groups, and organizations in civil society or NGOs)

- Build consensus on the kinds of issues that are on the policymaking agenda and communicate those issues to those who are responsible for policymaking

- Develop a clear set of activities and output as well as indicators to measure progress from the first dialogue to the last

- Establish strong, collaborative relationships

- Create a strong, cooperative group that can support the development of policy in local areas

- Consider constructive linkages between the policy dialogue and other dialogues addressing the long-term food security of the region

Paying more attention to the process and to building relationships than to outcomes and dialogue structure is also important because no single and unified approach exists that can be adopted for any context. Multistakeholder dialogues are nonlinear and iterative in nature. A dialogue does not start at point A and end at point Z, with the same agenda throughout. It is full of uncertainty, and its outcome is not predetermined but rather changes depending upon the interests of the stakeholders. Stakeholders have to manage the complexity of the issues as they move through the process. Thus they need to have contingent approaches that recognize institutional and political conditions and the opportunities and constraints these conditions may imply. Developing strong communication, information sharing, and trust among the participants will better enable them to withstand differences that emerge. The potential is present for governments in the Southern African Development Community (SADC) region and their development partners to expand and lengthen existing dialogues at the national and the regional levels and to initiate new ones. The experiences of these processes will teach us what they have achieved and how they can be made more effective.

Sharing Information and Building Awareness

The decisions of participants in multistakeholder dialogues and policymakers on the use and safety of agricultural biotechnology must be based on credible scientific information that all the stakeholders accept as valid. A key problem in the debate over biotechnology is the existence of false information and misrepresentations. In the absence of accurate information and the dialogues that help stakeholders to achieve consensus on it, conflicting claims arise that only make decisionmaking more difficult. More information on biotechnology, both for the dialogue members and for society as a whole, would build greater awareness and understanding of the issues and facilitate agreement on the issues and sound policymaking. Two general types of information would benefit the different stakeholders in southern Africa and the dialogues in which they engage: information on the technology itself and information on how the dialogue could increase awareness and participation and improve information sharing among its members.

Among other things, focusing on the process means engaging in a collective effort to obtain the information necessary to develop good policies and regulations. A dialogue at the national or the regional level in southern Africa should be informed on an ongoing basis by as much relevant information as possible on the major developments in agricultural biotechnology and their applications in the region. This should include information on the likelihood, frequency, magnitude, and distribution of the various outcomes from GM agriculture, and also information on the policy options for reducing the negative outcomes and enhancing the positive, based on the best available scientific knowledge and knowledge of local contextual features. To make decisions that society would accept, it will also be important for those engaged in a dialogue process to obtain and consider information on the social values attached to each of these outcomes by various groups, the level of uncertainty associated with various outcomes, the social values attached to that uncertainty, and the policy options for reducing or coping with the uncertainty. Greater awareness, dialogue, and consensus on alternative institutional and organizational arrangements for governing biotechnology are also needed. Working toward solutions will be easier if participants use a process of "joint fact-finding" to produce a common understanding of the likely effects, benefits, and costs associated with alternative policy options. Supplied with the available knowledge on the issues, eventually the dialogue process itself will generate information by monitoring research activities or policies implemented.

The governments in the SADC region will also need to support awareness building on biotechnology across the general population, because their people have a right to know how the technology might affect their lives, but misconceptions about it exist at all social levels. An informed society will influence national policy-making and research on the issue for the better. To disseminate information, civil society groups in the SADC countries and networks among them may be used. Countries with low levels of public awareness activities may be able to work together, as many of the issues and contexts for awareness building are regional in nature. Educating the population, particularly the poor, will bring benefits to the dialogue process, as it will help strengthen the capacity and knowledge base of farmers and consumers for participation in the process.

Awareness building can in fact be more successful if knowledge is gathered on the effective approaches that have been used to generate and share information. The dialogue could begin by collecting and examining what countries in the region and outside it are doing in terms of public awareness activities on biotechnology and then developing best practices and deciding how participation can be improved. It would also be possible for those involved in individual dialogue processes to form links with one another to share information on communication strategies and how national and regional networks and civil society and research organizations have disseminated their findings. What is particularly lacking is information on processes of policy formulation on biotechnology and the role of the different stakeholders in these processes. The understanding of the institutional and political context within which science and technology policy is made in Africa, especially with respect to biotechnology policy, is especially weak. Some 52 meetings on biotechnology were held in Africa in 2002, and a lot of information is already being gathered. Those participating in the dialogue could benefit from and add value by analyzing these processes and drawing lessons for themselves and others.

Investing in Research

The most critical information southern African stakeholders and policymakers need is on the benefits and risks that biotechnology would bring to their region, and only long-term scientific research can provide answers on these issues. But there is a dilemma here: short- and medium-term action is needed for food security in the region, but long-term research is needed, too. The ethical issue of the need to address the hunger that exists today cannot be avoided. However, there are currently knowledge gaps related to GM crops and biosafety, making uncertainties pervasive. A stakeholder dialogue can guide the research process and form a more effective link between the dialogue and policymaking. Because of their increased awareness of the potential dangers and benefits of the technology, policymakers are in a better position to see the need to develop necessary regulatory frameworks. All stakeholders, too, have different questions that they want answered. By taking these questions and finding ways to jointly frame them for the research community, dialogue participants can generate the information they need to reach consensus on policy measures.

As David Pelletier points out, some scientists in the biotechnology debate have been deciding how much and what type of uncertainty should be tolerated by society, and (together with regulators and politicians) discounting or misrepresenting these uncertainties in communications with the public. The appropriate role of scientists, especially those working in public research institutions, is to reduce the level of uncertainty through research and improve the methods available to test for adverse outcomes. Yet unfortunately research of this type has often been neglected in the case of agricultural biotechnology. In part this reflects the lower value researchers, their institutions, and funding agencies place on unintended consequences. Scientists in southern Africa can avoid this mistake. Indeed much more needs to be known, such as the nature of the relationships between GM crops and soils or the impacts of climatic conditions on ecological safety, which environmental scientists say is very important. And more information is needed about the whole range of food safety concerns related to GMOs under the conditions experienced by African populations, such as vulnerable health status and diets with very large shares of single commodities. Some of the main purposes of participating in the dialogue should be to guide, learn from, and provide feedback to research organizations in the region and internationally.

However, a dilemma the dialogue participants will face is that while the process is gradually moving forward there will be measures that they will have to adopt, or issues they will need to address rather urgently. These are issues regarding biosafety and trade issues that relate to GM crops and foods. Yet there appears to be consensus about the need to deal with these issues, whether out of a desire to protect the environment, farmers, or consumers; in response to the GM food aid controversy; or as a step in examining how national regulations can be harmonized with international agreements. If these issues are addressed within a dialogue, the resulting efforts and policies could be more successful.

Promoting Biosafety

One critical problem that was exposed in the debate over GM food aid is that the majority of countries in the SADC region lack the regulatory and scientific assessment structures necessary to take decisive steps on biotechnology. Only three countries in the region, namely Malawi, South Africa, and Zimbabwe, have legal mechanisms for biosafety. The rest are still at varying stages in the development of their biosafety systems. Most of the countries did not prioritize development of biosafety regulatory structures because of the low level of biotechnology research and development in their countries. If lessons from the 2002 regional food crisis are anything to go by, the countries in the region are best advised to put their regulatory and scientific monitoring mechanisms in place, because GM products may enter the region not from research efforts going on there, but instead from trade in such products developed elsewhere. The food aid controversy underlined the fact that in a globalized economy the development of biosafety regulations is not a luxury, but a necessity. For the long term, the SADC countries will benefit from the regulations created, as they will provide an enabling environment and monitoring mechanisms for biotechnology research and development and the use of GE products. A particular challenge to each country will be harmonizing regulations among their different public agencies, with other countries in the region, and with international agreements. Success in designing and implementing effective biosafety policy frameworks at the national and the regional levels will depend on national and regional commitment and cooperation, which a dialogue process can facilitate, as well as attention to the different country contexts and to capacity building.

In Chapter 1 Doreen Mnyulwa and Julius Mugwagwa inform us that opportunities exist for the SADC countries to collaborate, share information, and create synergies through dialogues. Given that three of the SADC countries already have biosafety systems, the experiences of these nations can be shared to allow for learning and adaptive implementation. That all the countries are signatories to the Cartagena Protocol could facilitate harmonization among the biosafety frameworks of the different countries for the transboundary movement of GMOs. Some of the goals of a dialogue should be as follows:

- To debate and come up with solutions as to how to harmonize regional policy on biosafety

- To link biotechnology and biosafety with trade policy

- To examine the missing links between national and regional policy approaches and determine which issues can be best addressed regionally versus nationally

In creating biosafety frameworks the stakeholders of the region will need to give attention to their respective economic, social, and cultural contexts. They would benefit from critically examining the dominant approaches to biosafety in the world, namely those of the European Union and the United States, the latter of which is used as a model in international development circles. However, these approaches are likely not entirely appropriate for the SADC countries. Whereas in the European Union modern biotechnology spurred the development of new regulations, in the United States scientists and regulators decided not to introduce new laws for biotechnology products but to rely on the country's existing regulatory structure. It is important that the southern African countries become very knowledgeable about the U.S. Food and Drug Administration's policies and their scientific, legal, and political bases so that they can engage in discussions and negotiations on biotechnology on a more equal footing.

The importance of developing biosafety frameworks that are attuned to the cultural food habits and economic and health conditions of southern Africa is illustrated by the U.S. experience. By not taking these considerations into account in making food safety determinations, U.S. agencies created a danger of announcing that GM crops are safe when they are not necessarily so for all populations. The population of southern Africa consumes unique foods, uses unique food processing methods, and relies on staple foods, such as maize, for the majority of their caloric intake. Furthermore, the high prevalence of morbidity, malnutrition, and compromised immunity due to HIV needs to be considered when testing GM products in the region. Contextual factors such as these will require greater attention in the future as GM foods with more complex changes come under development. An examination of how the scientific, legal, and political matters related to the new technology were addressed in the U.S. context holds lessons for southern African countries as they ponder the most appropriate institutional and procedural mechanisms for them to use to reach judgments, identify policy choices and trade-offs relevant to their region, and develop policies of their own. There is a clear need to balance benefits to human health and the environment with risks. People in the region need to feel safe and assured that their safety, health, and beliefs have been taken into account as far as possible before new forms of food products are introduced.

Key aspects of a biosafety framework should include the following:

- Legislative frameworks that include provisions to address trade-offs across public agencies in various sectors (e.g., agriculture vs. health vs. environment) and stakeholder groups (e.g., farmers vs. consumers)

- Clear criteria for selecting products to be submitted to regulation

- Unambiguous requirements for transparent state action and enforceable provisions for vigorous public involvement

- Rigorous risk assessment and management

- Communication with stakeholders on national biotechnology strategies and policies

Governments can use a number of specific measures to reduce the potential food safety risks of GM foods:

- Mandatory (rather than voluntary) premarket testing of new products

- Greater standardization of testing methods and decisionmaking criteria

- The use of newly emerging broad-spectrum profiling techniques to detect unintended compositional changes

- Consideration of the diverse contexts in which a given GM product may be consumed when developing, testing, labeling, and exporting or importing GM foods

In Chapter 5 Unesu Ushewokunze-Obatolu offers the following among several general recommendations for the creation of biosafety policies:

- Strategic action plans should be developed to realize the objectives set out to address selected policies.

- Member countries should be urged to design policies and actions that can be extended into regional and international arrangements.

- Member countries and the SADC should review their resource base to ensure that they can make effective commitments to allow biosafety processes to begin taking effect sustainably.

- Member countries and SADC should review existing biosafety mechanisms, infrastructure, and the human resource base to determine which functions can begin immediately and which can be phased in over time according to a schedule.

- Regional efforts to enhance biosafety research and testing should be promoted to reliably inform regulatory authorities and other regional decisionmaking structures in order to facilitate movements and trade involving GMOs.

- Investments should be made in systems for the retrieval and exchange of relevant information in order to establish national and regional biosafety information nodes for storage.

- The legislation and regulatory mechanisms adopted should be sufficiently flexible to account for the dynamism of biotechnology and biosafety and for their rapid development.

To develop biosafety regulatory frameworks, the countries in the region will require the necessary capacity in a number of areas. As a preliminary step, the governments and stakeholders can identify the capacity gaps. Improved skills and knowledge will be needed in the areas of scientific research, regulation, legal services, and policy. Based on the gaps, the actors can take decisions regarding the areas in which investments to close the gaps are needed immediately and the areas in which biosafety functions can be phased in once the capacity necessary for them exists. Capacity-strengthening strategies for biosafety will have to be prioritized and must be realistic. The countries of the region could conduct assessments and develop capabilities individually. However, they could also do so through regional cooperation, and given the differences among the countries in terms of biosafety development, there could be regional actions to coordinate cross-border capacity building. The SADC is well poised to provide leadership in this area and in others concerning biosafety development. Regional coordination of efforts for creating effective regulatory systems, including their harmonization, will also improve regional economic activity and food security.

Facilitating Trade

Divorcing biosafety from trade matters is difficult, because GM products constitute an increasing portion of exported and imported goods in the global economy. Indeed, in order to continue participating in world trade all southern African countries will have to develop biosafety policies that enable them to evaluate GM products entering the country for environmental and food safety. Trade in GM crops and food, which may play a significant role in food security, makes the formulation of biosafety regulations urgent.

Increased agricultural and food trade among the SADC countries is likely to bring benefits to all of them in the form of growth and food security. For this reason, harmonizing the biosafety regulations of the different countries would make sense. Given the similarities among many of the countries in terms of economy, ecology, and food habits, it would also not be difficult. However, the World Trade Organization (WTO) is putting pressure on countries to harmonize their policies with its regulations. Although making their policies compatible with regional and WTO standards would facilitate trade for these countries, each country should be able to establish regulations that meet its needs and goals.

Biosafety guidelines are vital for the southern African countries to enable them to decide whether they should receive GM products as imports or food aid. But they will be absolutely necessary if these countries wish to be among those in the world that are developing and exporting genetically engineered agricultural goods. In fact, fears have arisen that because the traditional exporting nations have adopted biotechnology, they will increase their exportable surplus, depress world prices, and make nonadopting importing producers, such as countries in Africa, less competitive. This would add to the problem for southern Africa's countries, particularly the poorer ones, of protected markets and subsidized farmers in developed countries. Yet the introduction of biotechnology provides an opportunity for developing countries to produce higher yields, lower their production costs, and source cheap agricultural exports. At the same time, the SADC countries may enjoy these benefits at the cost of reduced access to key markets, especially in Europe, where consumer sentiment against GMOs is likely to remain high well into the future. Preliminary questions countries of the region will have to ask are these: Which are the major traded commodities for which there are GM variants? Are these crops potential export crops for southern African countries? And how might the production of these crops affect exports to market of long-standing importance to the region?

Different consumer preferences in the world regarding GM foods—and, as discussed earlier, the environmental, food-habit, social, and health conditions in southern Africa—indicate that it would make the best sense for the SADC countries to develop biosafety and trade policies that suit their respective needs, despite pressure from the WTO to conform to its guidelines. In reality, the contention over the trade in and safety of GMOs has been caused by the lack of an international standard. For better or worse, this has given WTO member countries room to adopt trade-restrictive measures on GMOs. For example, the WTO recognizes environmental concerns, but thus far these concerns have not been tested in a legal dispute. Moreover, although the Cartagena Protocol on Biosafety, to which all the SADC countries have acceded, is an international agreement on procedures for the safe transboundary movement of GMOs, it is not clear whether the WTO will recognize the protocol's regulations. Finally, the WTO currently focuses on environmental safety. However, food safety is also a vital issue, and presently the regulations on GM foods in the WTO treaty remain undeveloped.

The harmonization and rationalization of national and regional policies on biotechnology and biosafety is a goal that the governments and other stakeholders in the countries of southern Africa should and can achieve. Harmonized legislation would facilitate the smooth movement and transit of GM material within the region, whether for commercial or noncommercial purposes. Clarifying national guidelines among the different ministries involved is a step that must actually be taken first. The SADC countries should harmonize their policies and procedures for standard setting and enforcement, risk assessment and management, prior informed consent, and information and documentation. At a minimum, the rationalized and harmonized policies should facilitate the approval and movement of products in the region.

The production of GM crops certainly has the potential to bring economic benefits to small farmers and food security to the SADC countries. But as Moono Mupotola reminds us in Chapter 7, it is not a panacea that will resolve the trade-related difficulties the region faces. If the area fails to address the export subsidies and protected markets in developed countries and their adverse effects on developing countries, little benefit will result. It is within the SADC's interests for member countries to act as a cohesive group and participate fully in areas of mutual interest during negotiations of international agreements, especially the WTO agreement. If they could influence the world trading system overall, the SADC countries would not have to rely solely on preferential market access opportunities alone.

Strengthening Capacity in Research, Policy Design, and Policy Implementation

For policymakers in southern Africa to possess the will to address biotechnology issues is the most important step. Following this they will need, in cooperation with the other stakeholders, to develop the requisite capacity in their countries in the areas of scientific research, policy design, and policy implementation, which will enable them to develop sound strategies for agriculture and for consumer and environmental safety.

Capacity is needed in several areas to develop and implement consistent biotechnology and biosafety strategies, policies, and regulatory systems. Core scientific capabilities and infrastructure are required for research on GM crops and, regarding biosafety, on biotechnology product evaluation, risk management, inspection, and monitoring. Equally important are competencies in managing the institutional processes that support these activities. Policy analysis and development capacity for biosafety, including trade issues, deserves attention, as these issues are relatively new and policy managers may not have the necessary backgrounds in them. Legal abilities in particular are lacking due to a shortage of legal professionals with an understanding of biotechnology. Biotechnology and biosafety know-how may be lacking in the officials in charge of regulations. The SADC lacks institutional capacity at both the national and the regional levels. One of the outcomes of this has been the region's failure to adopt appropriate time-bound performance indicators for its protocol ratification processes and programs.

Capacity strengthening for all the different areas and for the whole region will take time. The southern African governments should therefore ensure the provision of long-term funding for this goal. They will also need to prioritize the areas for capacity building based on their broader policies on biotechnology, biosafety, and trade and must have at least a degree of capacity for risk assessment and risk management.

Given the varying levels of capacity and resource endowment in individual SADC countries, structures and mechanisms for collaboration and the development of synergistic relationships should be developed to facilitate the pooling of resources across countries. The dialogue process can assist in the identification of capacity gaps and in the sharing of knowledge on experiences. Harmonization of policies across the region will also make regional efforts toward capacity development more manageable. Governments must develop strategic arrangements for technology transfer and expertise sharing with relevant private and nonprofit organizations both within the region and elsewhere in the world, taking care to clarify issues related to intellectual property rights and commercial confidentiality. In addition to regional bodies of the SADC and governmental organizations, NGOs can play a valuable role in strengthening national and regional capacities to make informed decisions on biotechnology. The aim should be self-sufficiency in all but the most specialized abilities. This would place the region on an even footing with the developed world in discussions and negotiations on biotechnology issues.

Developing a Broader Food Security and Poverty Alleviation Strategy

Adopting biotechnology for agricultural development, if done responsibly, can bring significant gains to the countries of southern Africa. But the specific role this technology will play in development and where the region will acquire the elements of this technology are issues that the governments of the region will need to clarify. The production of GM crops will be only one element of a broad set of strategies to achieve food security, poverty alleviation, and development, and the governments, other national stakeholders, and bodies at the regional level will have to consider a number of issues in deciding what part it will play in the region's broader biotechnology strategy and what other elements should be included, based on the benefits they expect to realize from this technology.

Should one of the countries that is presently not growing GM crops decide to do so, it will have to decide whether its own research institutions will develop the technologies or whether it will procure them from outside firms. If the technologies of multinational research companies are obtained, greater clarity in the policies of the southern African countries on intellectual property rights (IPRs) will be required. Although southern African countries have acceded to one or more regional or international agreements on IPRs, there is a lack of clear-cut policies on them in most of the countries in the region. Strong IPRs can provide the incentive private companies require to sell their technologies. As a result of the technologies, advocates of protective IPRs argue, a country can make advances in agricultural growth and food security. Although few African countries have the resources to develop their own large biotechnology programs, they could still benefit from the technologies of foreign firms.

Yet as Norah Olembo states in Chapter 6, in the southern African region there appears to be a lack of appreciation of the role of IPRs in development. Governments in the region therefore ought to clearly define the level of protection they want to provide for biotechnology innovations and consider conforming to the provisions of the Trade Related Aspects of Intellectual Property Rights agreement should they decide to procure technologies. For their own benefit, they will also need to decide on the desired extent and use of IPRs and determine the cost implications. There is a growing need for partnerships and collaboration among southern African institutions and multinationals in the area of technology transfer, which could enable research on crops important to the poor. But even these arrangements will require clarity on IPRs.

An alternative exists that allows countries in the region to develop legislation that protects the rights of farmers as well as indigenous knowledge and resources. In response to the International Union for the Protection of New Varieties of Plants agreement, in 2002 the Organization of African Unity published *The African Model Law* to protect the rights of local communities, farmers, and breeders and to regulate access to biological resources. The document was developed as a model for African countries to use to develop their own national laws. To date, though, no such laws have been enacted. IPRs should be coherent and should balance the rights of the innovators with those of the poor. They should also reflect the needs of the country and its development goals. Regardless of whether the governments of the region decide to develop technologies themselves or lease them from outside, the protection to be granted to breeders and to small farmers and resources in the country need to be well articulated. A dialogue process can help to bring the different stakeholders and the private firms together to ensure that IPRs do not conflict with the public interest.

As the southern African countries ponder whether to adopt biotechnology for food security and poverty alleviation, they will have to answer a number of questions, some more specific, others broad and fundamental. One set of questions relates to the opportunities for biotechnology and areas that require intervention. The countries of the region will need to determine individually, given their economies, what needs biotechnology can meet and specifically what crops should be targeted or what traits developed. As discussed earlier, genetic engineering technologies and the systems to ensure their safety need substantial financial investment and capacity, and countries are best advised to invest in areas in which they have sustainable competitive advantages or in areas that address their priority food security needs.

A related question is this: should the approach to adoption be reactionary in the sense that a country or a subregion should merely procure innovations developed elsewhere, or should the policy be a proactive one whereby the country or the subregion can produce technologies specially designed to meet the needs of subsistence farmers, consumer health, or HIV/AIDS-burdened areas with certain micronutrient deficiencies? If both strategies are pursued, where should the balance lie?

Governments and other stakeholders should ask a number of fundamental questions about the place biotechnology should have in the southern African countries' development strategies. One of the most pressing questions in the ongoing debate in the region is this: what will the technology, and all the investments required for it, contribute to food security? As one of the participants in the regional dialogue said, cotton is not going to solve the food insecurity problem. Investments in biotechnology will need to be considered in the context of national agricultural development and food security plans.

Several aspects of the southern African context need to be taken together in determining whether biotechnology has a role to play in development and precisely what positive effect it is expected to have. There is continued uncertainty about the possibility and seriousness of both food safety and environmental problems resulting from GM products. At the same time, food insecurity is a major problem in the region and will remain so. GM crops may help alleviate hunger and malnutrition, but it is not clear to what extent and how they will do so, especially if the underlying causes of these problems are not simultaneously addressed. Another question southern African governments have to answer is this: what policies do we want to pursue given these uncertainties and conditions?

The response to this question will depend to a significant degree on the relative importance to policymakers of reducing household food insecurity and malnutrition, especially among vulnerable groups such as women and children, and reducing sickness. But both GM-inclusive and non-GM policy options are available for achieving each of these goals. What are the potential benefits, risks, and costs associated with the policy options in each group? Are the GM-related policies superior generally? Can GM agriculture contribute significantly to improving food security and nutrition in southern Africa without creating unacceptable risks to food safety and the environment? These are questions that the governments, farmers, consumers, and private-sector and other stakeholders in the region will have to address together.

A view that many critics of biotechnology have expressed is that it is a technological solution advanced to solve problems that at root have political and economic causes. Non-GM policies to eradicate hunger and malnutrition have been implemented and shown success when they were designed to suit local contexts, were well managed, and received the requisite levels of political, institutional, and economic support. So is there a need for GM adoption? Moreover, with the introduction of biotechnology, these basic and necessary policies may be neglected. It is also being increasingly recognized that food security depends on the broader foundation of good governance, peace, rule of law, respect for human rights, and equity in development. Even if GM technologies are applied, it is likely that if they are to ultimately have a positive impact on malnutrition and food insecurity it will be necessary to continue and even expand the "conventional" programs that have been implemented to these ends and to improve governance. Some examples illustrate the need for programs for nutrition, health, employment and income generation, education, safety nets, legal rights, and other goods to accompany the adoption of biotechnology. For instance, iron and pro–vitamin A (beta carotene) in plants has very low bioavailability, so enhanced levels of these nutrients in GM foods may have little or no impact unless the quality of overall diets is also improved. Improved household food security through GM agriculture—if achieved—will not reduce child malnutrition unless governments also invest in programs for child health, child care, and child feeding, all of which women have difficulty providing due to their own poor health, nutritional status, and knowledge, as well as time demands. Another question for the countries of southern Africa to ask is this: if there is weak commitment to provide the types of programs and the quality of governance on which GM adoption will depend to generate benefits, will it make sense to pursue the application of biotechnology for food security and poverty alleviation?

Creating Sustainable Financing Mechanisms

Concerted efforts to formulate and implement biosafety strategies, policies, and regulatory systems require reliable and sustainable streams of financial resources, especially to meet the heavy burden of capacity strengthening. If the SADC countries choose to develop innovations in biotechnology—and some are already doing so—they will also need to invest in research over a long time frame and in a steady manner. While multilateral and bilateral donors are likely to be willing to support these actions to promote national and regional ownership and control of the biotechnology agenda, the nations of the regions must also be willing to commit their own resources. They can do so either individually or collectively via the SADC. Obtaining donor support and allocating resources effectively will hinge on clearly defining and gaining broad acceptance on the national and regional needs and priorities. The countries can take the important first step of identifying these needs and priorities under the aegis of the SADC. A collective effort could bring greater efficiency and more rapid outcomes.

The Food, Agriculture, and Natural Resources Policy Analysis Network–International Food Policy Research Institute initiative on agricultural biotechnology is based on a vision of catalyzing an expanded and sustained regionwide dialogue among the national governments, regional bodies, organizations of farmers, the poor and consumers, those representing the environment, and the international agricultural research and donor community on the future of the technology and of biosafety in southern Africa. It is hoped that this multistakeholder process will also generate cooperative action on the part of the members to take the necessary steps for ensuring the safety of the region's population and environment and for responsibly pursuing biotechnology-led agricultural development. A dialogue process will assist the countries of the region in assessing the benefits and risks of biotechnology for their respective cultures and the environment as each decides which direction is best for its population.

To continue and develop the dialogue, those promoting the initiative envisage an integrated series of forums on biotechnology, agriculture, and food security in southern Africa that are carefully facilitated and highly participatory, involving a significant number of high-level policymakers, senior representatives of a range of stakeholder agencies, and respected scientists. Interlinked roundtable gatherings are planned to take place over several months. Bringing different views to the table for deliberation and information sharing has the best chance of building consensus, which could then lead to the collaborative planning, implementation, and evaluation of various activities.

But deepening the dialogue and involving more parties in it will not be without its challenges. It will experience bumps at times, strong conflicts among members, and possibly dissolution due to this conflict, a lack of interest, a shortage of resources, or other factors. What will enable it to surmount these obstacles and continue will be a focus of the stakeholders and facilitating organizations on the process. Building trust among the members, maintaining communication, exchanging knowledge, and being open to revisions of old views will not only help the dialogue last, but will also be more productive.

Naturally the parties in this dialogue will also have to possess an interest in and make a long-term commitment to moving forward. If they lack the resources or capacity to participate, it will be necessary for them to acquire these, which may be done through the dialogue as an institutional structure. Some who are skeptical about multistakeholder dialogue processes are present who believe such approaches are unnecessary for action to be taken on biotechnology. However, in light of the conflict that has existed on the issue and that is likely to grow more intense in the future if honest deliberation does not take place, the question is this: what will the outcomes be for the countries of the region if a dialogue is not established?

Initially it may seem that the multistakeholder process is taking time. The process may also seem to be too precautionary, that is, antiscientific, antitechnological, and insensitive to the poor. But addressing biotechnology in this way will bring all the concerned parties on board and get them moving together in some direction on the numerous issues. A process of this kind will also adopt a broader view of the issues and of development, food security, and poverty alleviation, and will make sure that the knowledge provided members on the benefits and risks of the technology is reliable. By working carefully and collectively, the process will also be more open, transparent, inclusive and accountable, and sensitive to the normative dimensions of the issues so critical to the participants.

Reference

Nuffield Council. 1999. *Genetically modified crops: Ethical and social issues.* London.

Appendix A

Workshop Proceedings for the FANRPAN-IFPRI Regional Policy Dialogue on Biotechnology, Agriculture, and Food Security in Southern Africa

Jenna Kryszczun and Steven Were Omamo

Meeting location: Senators Hotel, Caesars Gauteng, Johannesburg, South Africa
Meeting date: April 25–26, 2003
Meeting moderator: Dr. John Mugabe, New Partnership for Africa's Development Science and Technology Forum

DAY 1

Welcome and Introductions

Presentation: Dr. Tobias Takavarasha, Food, Agriculture, and Natural Resources Policy Analysis Network (FANRPAN)

Dr. Takavarasha opened the meeting, noting that the initiative was the result of many interactions with the International Food Policy Research Institute (IFPRI), regional experts on biotechnology, and government representatives. He expressed his hope that the dialogue would contribute to the development of proposals and recommendations to address the critical subject dealt with in many high-level meetings—issues related to genetically modified organisms (GMOs). Dr. Takavarasha further noted that one of FANRPAN's central roles is to facilitate policy dialogue such as this one within the region.

Meeting Moderator: Dr. John Mugabe, New Partnership for Africa's Development (NEPAD) Science and Technology Forum

Dr. Mugabe clarified his own role in the meeting, namely to facilitate dialogue. He welcomed Dr. Joachim von Braun, IFPRI's director general, to outline IFPRI's expectations for the dialogue. Each participant would also have an opportunity to express his or her expectations for the two days of dialogue.

Objectives, Expectations, and Ground Rules

Presentation: Prof. Joachim von Braun, IFPRI

Dr. von Braun indicated that IFPRI's role in this event and in this process would be that of a partner to African organizations.

Both IFPRI and FANRPAN view it as very important to put this theme of biotechnology policies for the southern Africa region on the agenda, because it is felt that this is one of three key issues that need to be addressed (the others being food security and land issues). IFPRI is addressing this theme because it is a global issue as it is not only a hot issue here in the southern African region.

IFPRI's perspective on biotechnology is that developing countries must make their own informed decisions. Biotechnology may become a key driving force for economic and social development. The questions are these: For whom? For what? What regulations should be put in place? How can biotechnology be made to work for the poor? And how can technological policy options be employed that are environmentally sustainable, efficient, and effective?

In discussions in various parts of the world regarding the state of affairs in agricultural policymaking (in Africa, in Asia, in Latin America), this issue is on the agenda of policymakers and continues to irritate consumers and producers. Opportunities are recognized, but potential risks are acknowledged as well. The perspective differs by region, but in all regions key uncertainties and controversies in biotechnology policies remain unresolved. IFPRI has the opportunity and challenge to facilitate learning around the world. The dialogue should therefore be inclusive and largely driven by actors in the region.

The process of this policy dialogue aims to raise awareness, promote dialogue elsewhere, and drive toward consensus building. This meeting is embedded in this larger process. The short-term aim is decisionmaking support. The ultimate aim is consistent institutions and policies that govern biotechnology policy and related capacity building based on increased understanding and greater awareness and consensus on simple policy trade-offs between benefits, risks, and sustainability.

The key questions to be addressed at this meeting are these:

- What are the major issues of debate and dispute dynamics around agricultural biotechnology in the southern African context?

- Who are the relevant stakeholders? Many of them sit at the table, but not all of them.

- What questions might be suitable for joint fact-finding and follow-up?

- What are the constraints—financial, institutional, legal, and technological—under which we are operating? It takes resources to undertake technology assessment dialogues. It takes time, commitment, and money.

- What sort of commitments can be made by this group for follow-up actions?

The expectations and ground rules are easily spelled out:

- Participation is to be active, open, and fearless.

- There is to be no privileged position a priori.

- Every opinion matters.

- Disagreements are to be open and respectful.

- A constructive, forward-looking perspective is to be adopted.

The aim is to have not a general debate on biotechnology, but rather one on legal and administrative decisions and processes governing biotechnology, and on how to engage with them fruitfully. To what extent are these decisions and processes driven by the international legal environment (e.g., biosafety, the Convention on Biological Diversity, and the upcoming World Trade Organization [WTO] negotiations), and how can this initiative feed into them? This is a first step that could grow into a global process.

Meeting Moderator: Dr. John Mugabe, NEPAD Science and Technology Forum
Dr. Mugabe began by noting that good policies are formulated just before cabinet meetings and after cabinet meetings. Policies are in many cases developed in very

informal settings. He encouraged the group to strive for informality. He went on to make the following points.

There are many policy questions that must be addressed, and the range of controversial issues that governments think about is growing. The group may therefore wish to reflect on those issues, without necessarily aiming to arrive at consensus on all or any of them. The group might develop a typology of issues that regional governments and stakeholder groups are facing to see if biotechnology will be appropriately used for human development. The aim is not to invest a good part of our energies in debating the issues, but rather to reflect on what the issues are that need to be high on the political agenda and that need more research.

Some key questions that must be considered are these:

- Under what conditions are we going to engage various stakeholders?

- What constituencies will strive toward consensus, and what constituencies do we need to bring to the consensus-making process?

- It is very clear to those of us who talk about biotechnology every day that the debate in this region is still confined to very small, isolated groups. How do we go about building constituencies?

- What is the appropriate locus?

- Who is going to develop the biotechnology policies at the subregional and regional levels: ministries of agriculture? ministries of health?

- Given that the issues are wide-ranging, who should be at the policymaking table?

The aim is not to develop consensus, but rather to reach agreement on the nature of the process or processes that our countries and our region need to use to move toward consensus. Specifically, what kinds of process or processes are going to be required to ensure discussion of controversial issues? The idea is to reflect on the types of processes that have been used by other actors, as there are many groups already investing in policy development (at national and regional levels). Whom should we be seeking to influence, and how can we ensure that one ministry will not develop a policy that is in conflict with a policy developed by another ministry?

Dr. Mugabe then invited participants to raise any issues and offer their expectations. What did they want to get out of this dialogue? Were their expectations different from IFPRI's and FANRPAN's expectations, or was there a convergence?

The answers were provided in a moderated plenary session.

Moderated Plenary Discussion

Dialogue participants raised several expectations of the meeting, including the following:

- To debate and come up with some solutions on how there can be a harmonized regional policy on issues such as biosafety or the benefits of biotechnology

- To have a learning experience

- To link biotechnology with trade policy

- To build consensus on the kinds of issues that are on the policymaking agenda and to communicate those issues to those who are responsible for policy

- To arrive at consensus on the process and bring those that are not involved in this dialogue to the process, particularly farmers, consumer groups, and civil-society organizations or nongovernmental organizations (NGOs)

- To develop a set of clear activities and output as well as indicators to measure progress from the first dialogue to the last

- To establish strong, collaborative relationships

- To establish a strong, collaborative group that can support the building of policy in localized areas

- To find one or two items on which to focus action

- To learn from others at the meeting about effective approaches and processes (i.e., how to build confidence in the application of these technologies)

- To consider constructive linkages between this policy dialogue and other dialogues addressing the long-term food security of the region

- To explicitly state common interests, such as a bountiful and nutritious food supply that is environmentally sustainable

- To further understand how the policy arena and the research agenda interact and to examine the role of the public sector in the process

- To consider how to balance the government's need to face short-term pressing challenges (current food needs) with long-term issues, such as biotechnology

- To share experiences in terms of difficulties and successes in biosafety and biotechnology

- To examine missing links between national policies and regional policy approaches and to determine which issues are best addressed regionally or subregionally vs. nationally

- To hear more about the expectations, approaches, and role of the SADC Advisory Committee on Biotechnology

This question was also raised: Since this is seen as a long-term strategy fitting into other strategies, what are IFPRI's and FANRPAN's visions of how this first dialogue will fit into other processes? How will they move the outcomes of this meeting forward?

Dr. Takavarasha indicated that FANRPAN's role is facilitative in terms of communicating key policy findings to policy decisionmakers, whether that information is obtained through policy dialogues, publications, or research undertaken by policy researchers in the region. FANRPAN is trying to fill the gap that exists between what the policy researchers recommend in their findings and what policy decisionmakers are able to implement (specifically in the area of biotechnology, in addition to other areas such as land, food security, trade, and strategic food reserves, among others). He noted that biotechnology is becoming an important factor in food security and in trade, and yet the level of understanding by key decisionmakers is an issue that needs more debate. Dr. Takavarasha referred to the participants' mention of harmonization in terms of understanding, in terms of biosafety regulations, in terms of anticipated benefits of biotechnology, and in terms of effects of trade, and noted that the these are issues that need to be brought to the agenda in order to better equip policymakers. He explained that FANRPAN seeks to acquire as much information as possible so that, as a network, they can contribute by passing it on to policymakers quickly. There is a need for frequent meetings in order to address the information gap. One way to maximize the use of such meetings is to ensure that FANRPAN strengthens its linkages and dialogue with key policymakers, and another is to ensure that meetings that bring people together become as effective as possible.

Dr. von Braun relayed what influenced his thinking on the need for such dialogues, especially in this region. It was in his first week as director general of IFPRI in September 2002 of that he was visited in Washington by a delegation from

Zambia. That delegation was sent by their president, and asked a whole host of questions related to GMOs and biosafety, food safety, and trade and food culture–related issues. Dr. von Braun noted that it became pretty clear after three hours of meeting and exchanging perceptions that there was an urgent need of a better-informed debate. He continued, remarking that in looking at the diversity in the southern African region he noticed that the contrast in the area of biotechnology could not be starker between perceptions and level of use and the expectations from the technology. He noted that the issue raised about whether a subregional or a national approach may be more advantageous than a regional approach is a very important one, one that requires debate and analysis. However, Dr. von Braun questioned whether a national approach would be beneficial in the long run, because the subregion could benefit immensely from a coherent agricultural, biotechnology, trade, and rural development policy.

In response to the question posed about the vision brought to this dialogue, Dr. von Braun indicated that the dialogue helps in joint agenda setting, with milestones and goals to be achieved in the area of improving policy implementation of biotechnology policy. This can be structured in four areas:

- The agenda related to consumption and consumer benefits

- Concerns with environmental safety

- The area of trade, including intellectual property rights

- Issues surrounding politics, culture, and perception

This final area cannot be left off the table because, Dr. von Braun explained, for his visitors from Zambia this was at the top of the agenda (issues such as ethics, food culture perceptions, foreign policy, diplomatic complications, etc.). These issues do not lend themselves to the simplification of the science-based vs. value-based debate. Dr. von Braun indicated that he would be interested in being better informed about what is driving this fourth component, which is influencing the decisionmaking on biotechnology. He noted that these issues are very critical for the final outcomes of coherent food and agriculture biotechnology–related policies.

Dr. Mugabe noted that participants have common expectations of the dialogue. He offered two important points:

- The dialogue must add value to ongoing processes.

- The dialogue should be treated as a learning process; participants are here to exchange information.

Agricultural and Biotechnology and GMOs in Southern Africa: A Regional Synthesis

Presentation: Julius Mugwagwa, Biotechnology Trust of Zimbabwe
Key points made during the presentation were these:

- This overview is based on reports of various stakeholders and is intended to ensure that we do not start from scratch or reinvent issues, but start from the same level.

- Biotechnology has been around a long time, since the times when people started domesticating animals and plants. Only recently did a Hungarian engineer coin the term "biotechnology," but we have had these technologies for a long time. We are all aware of the work that has taken place in the late 20th century: the technology between the two world wars (penicillin); in the mid-1950s the deciphering of the structure of DNA, and further developments to do with the ability to cut DNA and hereditary materials, using enzymes; in the early 1970s the discovery or the invention of the ability to multiply genes, DNA; and in the 1990s cloning (Dolly the sheep). Now we have a lot of GM products on the market. People are talking about genomics and are trying to understand a whole sequence of genes in the human body. We are talking about a technology that involves the use of biological organisms in the production of goods and services. While biotechnology, working with genetic material, is generally accepted in medicine, it is controversial in other sectors, including agriculture.

- There are great differences among the countries of southern Africa regarding the scientific activity they have undertaken in biotechnology.
  - Agriculture is the main area where biotechnology is being applied in southern Africa, mainly in the dimension of crop improvement. A few countries are employing genetic engineering techniques, but these are only in the laboratory; only South Africa does so commercially.
  - All countries are employing tissue culture techniques, and most have invested in biological nitrogen fixation. Fermentation technology, marker-assisted selection, artificial insemination and embryo transfer, molecular diagnostics and molecular markers, and genetic modification are also being widely employed. Considerable training in these techniques is underway in the region.
  - South Africa is really active in almost all techniques. There are a lot of universities and institutions, and the agricultural research center is quite

involved. A number of products are on the market, for example *Bacillus thuringiensis* (*Bt*) maize and a cotton variety. Mauritius already has a GM sugar cane variety ready for release and is awaiting adoption of a biosafety framework in order for it to be released. Tanzania is mainly doing research work, but training is also really implicit.
    ○ Little information is available for Angola, the Democratic Republic of the Congo, and the Seychelles.

- The SADC countries also vary with regard to their development of regulatory frameworks. Most countries currently do not have biotechnology policies in place. The development of biosafety systems is needed to manage or to ensure the safe development and application of biotechnology. Based on a study that was conducted in 2001, countries can be placed in three broad categories:
    ○ Those that have legally binding frameworks in place already—Malawi, South Africa, and Zimbabwe (further clarification is needed on Namibia)
    ○ Those that have draft legislation—Mauritius, Namibia (waiting for clarification), and Zambia
    ○ Countries that are still in the initial stages, with very preliminary guidelines —Angola, Botswana, Lesotho, Mozambique, Swaziland, the Seychelles, and Tanzania
  Little information is available on the Democratic Republic of the Congo.

- There are great differences among policymakers in the region in awareness about scientific issues and specific policy details.
    ○ Countries were at different levels of preparedness to handle GM issues during the food crises brought on by the 2001–02 drought. There was limited awareness of biotechnology issues across the region, as evidenced by the debate. In some cases the debate was informed or driven by emotions, or it was subject to the big divide between the United States and Europe.
    ○ The awareness varies from high to low, and this depends on the category of stakeholder to which one is referring. Life scientists are highly conversant and quite aware when compared to other scientists and other people. A number of dissemination activities and awareness-raising activities are ongoing in some countries—for instance, those convened and coordinated by AfricaBio, Biowatch, and Safe Age in South Africa; by the Biotechnology Trust, the Regional Agricultural and Environmental Network, and the Biotechnology Association in Zimbabwe; and by the National Institute for Scientific and Industrial Research, the Biosafety Committee, and the National Biotechnology Alliance in Zambia. Also important are regional efforts by the

Southern Africa Regional Biosafety Program, Consumers International, the Biotechnology Trust of Africa, the African Center for Technology Studies, the African Biotechnology Stakeholder Forum, NEPAD, and the United Nations Environment Program (UNEP).

- Increased use of GMOs in the region is contingent on policy development.
  - There is a strong correlation between the state of policy development in a country and the level of use of GM techniques in those countries. Those countries that are active in the use of these techniques are also the countries that have legislation—for example, Malawi, South Africa, and Zimbabwe. Mauritius, Namibia, and Zambia are almost there.
  - Some of the challenges that are faced in trying to engage the public in this biotechnology debate are these:
    - Commercial confidentiality
    - Costs of various levels of participation
    - The interface between farming systems and social/cultural factors
    - External influences
    - Interactions among food aid, politics, science, and regulations
  - Meeting these challenges implies progress in the following areas:
    - Identifying regional needs and priorities
    - Building scientific and regulatory capacity
    - Creating an enabling environment for research and use of products
    - Promoting regional approaches to biotechnology issues

Moderated Plenary Discussion

The Chair highlighted some key points emerging from the presentation:

Technology investment. We are not dealing with a simple technique, but a system of techniques and a growing body of science. A distinction may need to be made between the techniques and the products.

Public awareness. Public awareness is important. Is it really crucial for a farmer to know what biotechnology (genomics) is or to know the content of the products? What pieces of information need to be provided to civil society and laypersons? Who raises awareness?

Policy formulation and research and development. As indicated in the presentation, those countries that have been able to develop policies and biosafety regimes have seen increasing investment in research and development. Do all countries require policies today? Do certain countries need to worry about research and development before they move into policy development? As countries build scientific capacity and new processes and products, policy questions may become critical. Is there a correlation between policy development and investment in research and development? Should countries wait to invest until they have a better sense of what biotechnology is?

Trade issues. Should countries be worried about what the impact of these technologies will be on their trade activities, whether in fact adoption of a particular product will undermine their trade relations? The presentation also alluded to the fact that biotechnology and GMO questions are increasingly moving into foreign policy domains. We need to think about how we influence foreign policy, as opposed to leaving discussions of policy development in biotechnology within the spheres of countries.

Intellectual property protection. It would be useful to better understand the content of the various biosafety frameworks, to which extent they are addressing biotechnology, and to which extent are they biased.

National, subregional, and regional agendas. It was clear that we are seeing a growing number of initiatives in biotechnology, and this increase could be a source of potential tension and conflicts. To what extent do they undermine our efforts as a subregion to reach consensus? What are the vested interests of some of these groups (which have clearly formed agendas)? We need to understand those agendas in order to bring the groups to the policymaking process.

EU and U.S. biotechnology policy. Is there a unified EU policy on biotechnology, given the nature of the investment each of the countries is making? Is there a U.S. policy? The two regions tend to be treated as if one has a more homogenous, uniform policy of pro-biotechnology and the other has a more formed policy of anti-biotechnology.

Food culture. In this subregion, is biotechnology not accepted because food is so part and parcel of our cultures that we do not want to taint food products by modifying them?

The majority of the discussion focused on issues surrounding regulatory systems and biosafety policy, protection of traditional indigenous seed and plant varieties, arguments for increasing investment in biotechnology research and human resource capacity, and exchange of information, among other items.

Participants raised the following points regarding regulatory systems and biosafety policies:

- If people are aware of their rights as citizens, they can monitor the regulatory mechanisms in place so there are no violations.

- If a country can police itself, how can the point be made that countries need each other? If a country knows that the regulatory framework and infrastructure are not there, can it police itself to control crossborder movement of GM products?

- The regulatory system in South Africa, which has been in place for almost four years, originally began with the UN debates on weapons of mass destruction and concerns about biosecurity. The debate evolved to focus on GMOs, and it soon became clear that there was a need to set up a system whereby people could apply for permits in order to operate. However, as time went on, the government discovered problems with compliance and crossborder exchange of seeds, which often has a cultural dimension. Questions arose, such as how to ensure compliance and how to make neighboring governments aware of violations. There is a need to find ways of handling the products and the seed. It has taken too long to bring a harmonized regulatory system into existence.

- Tanzania has a fairly advanced draft document on biosafety policy, and some scientists are asking why it cannot be used on an interim basis, because they do not want to stop at the field trial level, but want to move on to commercial production.

- Each country must first develop its own regulation policy, and then that policy can be harmonized with the policies of its neighbors.

- Governments have many demands on their resources, and biotechnology policy regulations are costly and compete with other government functions. How can these costs be cut without jeopardizing safety, efficiency, equity, and other considerations? In an environment in which government has deficiencies in implementing, regulating, and enforcing in the interests of farmers, consumers, and traders, it is a tempting alternative to shift the burden onto the private sector. This is a difficult proposition because of legal responsibility and accountability, but perhaps the roles of government vs. the private sector could be clarified in terms of shouldering responsibilities.

- Unlike research, policy is a very messy business, and events on the ground, where decisions are often made in an unclear policy environment, often overtake the measures taken at the policy level. Can the question of biosafety be considered in the context of the trade-regulatory environment?

- Most developing countries insist that farmers enjoy farmers' rights benefits, which allow for exchange of seeds whether across borders or within districts. The minute that right is denied, smallholder farmers' ability to successfully use GM crops to their benefit is precluded.

- For some time, farmers are going to share their seed; that way technology can benefit those who need it most. This ongoing system should be supported, and scientists may wish to look at these options to bring everyone on board in terms of benefit sharing.

- What countries have signed the biosafety protocol? The protocol should make it easy for countries to import GM products. South Africa has made the decision to sign, but must ratify it in order to accede. Many countries have not necessarily signed, but have declared their intention and are in the process of doing so. It is one of the criteria a country must meet to access funds for capacity building for biosafety.

- Biotechnology companies want to be seen as adhering to the law, and will not introduce seed into a country that does not have biosafety legislation. If the company discovers that seed has crossed the border, they will write a letter to the government indicating that it is not with their consent. However, it is inevitable, especially when countries have been exchanging germ plasm for years. This is why a biosafety system must be developed quickly.

- To what extent does commercial confidentiality exist? How readily is the information available to users, and does the information need to be known fully by all stakeholders? To regulators of countries, there is full disclosure; they demand to know everything, but they are bound by confidentiality. Users of the technology may be provided information on the function, benefits, efficacy, and scientific rigor, but not specific details on how the product is made, which could be commercially detrimental to the company. Increasingly, companies find that the more transparency and disclosure there is, the more acceptable the technology will be, given the surrounding controversy.

- The advantage of the regulator is that he or she is able to see what information companies claim to be confidential, and whether another company has claimed the same information. The regulators also try to allow public research institutions to obtain a level of access to resources in terms of benefits or the proprietary nature of some of the technologies in the hope that the information will filter down to the users.

- The regulatory systems existing in North America and Europe are being strongly revisited by consumers, industry, and policymakers and seem deficient in areas such as biosafety, food safety, accountability, and the responsibility of various actors. Southern Africa can learn from these experiences and avoid mistakes.

Another key issue is the protection and support of traditional indigenous seeds and plant varieties. Participants raised the following issues:

- There is a need to take particular care in promoting or supporting existing efforts to conserve what is indigenous to the region. Many crops are not commercially valuable but are of immense value to communities and farmers. The main challenge is to build capacity to be able to categorize and reference them for future or continued use.

- It is not that we should necessarily refrain from genetically modifying these crops, but we should know what they are before we replace them.

- To what extent have efforts been made to catalogue and patent traditional seed varieties? There is a great movement among countries with biodiversity to categorize and add the varieties to databases to ensure that patents on them will be stopped if attempted.

- Countries should also consider patenting plant varieties under the Trade Related Aspects of Intellectual Property Rights agreement. Databases to catalogue traditional plants can protect genetic resources from bioprospecting. It is crucial to introduce plant variety rights side by side with patent laws as a way to increase and protect farmers' rights.

- The *African Model Law* on new plant variety rights and farmers' rights, developed two years ago, has not been used much. It provides various options, and not all have to be taken on board.

- Although traditional seeds do not qualify for protection (because they are not new), companies would not be interested in them for the same reason—because they would not be able to patent them. However, elements from the seeds can be patented; they do not have to be protected in their original form. African countries have a lot of valuable material and must put in place mechanisms of accessing their plants, so they will not be left open for anyone to benefit from without returns to the people who have nurtured them for so many years.

A participant asked this question: if a ministry of agriculture needs to convince the minister of finance or parliament to pass an incremental budget to deal with biotechnology capacity building in research laboratories and human resources, what arguments could they use? Others responded with the following comments:

- An economy cannot survive in isolation; this technology should be obtained in terms of regional trade.

- If a country does not invest in this new technology, their environmental and food security will be undermined.

- If the country's economy is going to be competitive internationally, there must be some indigenous residual biotechnology.

- This technology will improve the well-being of the rural sector.

- Countries should be urged to build up indigenous laboratories and capacity to avoid putting themselves in negative power relations.

- This is not an outlandish technology of the West or one that is in the hands of multinationals. Indigenous institutions and our own scientists are working on this.

- Look at what the countries around the world with the biggest food security and population problems are doing. China and India are investing quite a lot in biotechnology, and would not do so if they were not receiving benefits. What benefits have they accrued from investing in biotechnology?

- Although few African countries will have the resources to develop their own large biotechnology programs, they are still able to benefit from the technology and should invest in regulatory frameworks and research in order to facilitate intelligent borrowing.

The group discussed the need to continually update the useful tables developed by Dr. Mugwagwa. Some corrections and additions were suggested, including these:

- Tanzania has made great strides over the past few years. Programs in microbiology and environmental and industrial technology have begun. A bachelor of science program in biotechnology is being started around issues of crops, agriculture, and medicine. Human resource capacity is missing but will improve.

- It was suggested that presentations be made about the extent of biotechnology research vis-à-vis the ongoing agricultural research within countries so that they can appreciate the relationship between the two.

- It was also suggested that comparisons be made of the type of research undertaken with the problems the country has and how biotechnology can help.

Several other issues were discussed among participants, and the following remarks were made:

- The question "Do we need this technology?" may be simply answered "Yes" in the scientific arena, but in many other arenas there are still many unresolved concerns. The debate must be as inclusive as possible, with all sectors involved.

- Efforts should be made to have common conferences with both extremes of opinions represented; perhaps the debate will then move forward more quickly.

- When both sides are represented, the outcome is rarely positive. The only way to engage is to provide information on the ground and correct the misinformation that has been provided to consumers by those campaigning against the technology so the consumers can make their own decisions.

- A central issue is networking and communication. Information needs to be disseminated. It would be useful to share the experiences of national and regional networks and civil-society, advocacy, and research organizations to see how information can be effectively packaged. The generation of the right information is also important.

- Countries with low levels of public awareness activities may be able to work together, as awareness issues go across borders. It is suggested that a working party be formed on how to create synergies to work on communications activities across borders.

- How do we enable those with their PhDs to put what they have learned into practice?

- Some of the challenges of this technology are due to inequities and the fact that some people are not able to take advantage of development. Biotechnology has moved beyond the natural sciences to the level of genomics and bioinformatics, and the ability to manipulate genes and develop a product is now closely linked with information and communications technology, to which not all have access. This has implications for intellectual property rights (IPR) issues (who owns what genes?) and for the modes of production in our society.

- How can biotechnology research be viewed as a long-term development strategy in terms of overall development strategies (PRSPs, national development strategies)?

- In South Africa, it was not until a national biotechnology strategy was developed with its own research priorities that the ministry of finance was approached for funding.

- When dealing with populations that are 70 percent rural and 70 percent below the poverty line, it is critical to ask this: What does this technology mean to a country that is trying to feed its population? If introduced, will the technology speak to that priority?

- What about risks and uncertainties? Within the region it would not be difficult to convince someone in a policymaking position that biotechnology research and testing is important, especially given the more frequent droughts. As of now, farmers use inputs such as fertilizers, pesticides, and herbicides, which are toxic and which are governed by rules to ensure that they are absent from food, just to achieve a successful harvest. It is a question of weighing costs and benefits.

The Chair closed the session, indicating that later discussions would not focus so much on the issues themselves, but on agreeing on a set of policy issues common to the countries in the subregion that the dialogue could address, and the process by which the dialogue would do so.

Dealing with Complex Public Disputes: Multistakeholder Approaches, Negotiation, and the Practice of Consensus Building

Presentation: Ms. Michele Ferenz, Consensus Building Institute, Cambridge, MA, USA

Key points made during the presentation were these:

- Conflict resolution through policy dialogue
  - Nobody likes conflicts; they are long, costly, and painful, and a lot of people wonder whether negotiations are worth their while.
  - One negotiation concept that is useful here is called the best alternative to a negotiated agreement. It is based on the idea that the only reason participants would want to enter a policy dialogue or a negotiation (a policy dialogue is an ongoing negotiation) is because they have decided that it is the best way to achieve their goals.
  - A person will not enter a policy dialogue or negotiation because it is fashionable or the right thing to do.
  - If stakeholders think the best way to achieve their objectives is by not engaging in dialogue, the whole discussion is moot. They either think they have the power in different arenas and the world will eventually recognize they are right or they will use other avenues of influence.
  - A complete consensus will never be reached on a complex policy question, because there are people whose whole identity revolves around being against an issue. However, one should not disengage from a dialogue for that reason. There is value to interaction and dialogue with stakeholders whose opinions are more open.
  - Policy dialogues have been held around very emotional, complex issues. For instance, the World Commission on Dams held a global multistakeholder dialogue for two years involving thousands of people and a lot of resources. It is the ideal example.
  - The question for us is this: What, given the constraints of this region, can we achieve moving forward in our process?
  - The concept paper explains that we want to construct a policy dialogue involving lots of different stakeholders. The objective of this presentation is to address the following questions:
    - What are policy dialogues?
    - Why do we have them?
    - What are the gaps they fill?
    - What are the difficulties with them that are far from being resolved?

- A key issue already raised at this meeting is the concern about false information and misrepresentations. Some of the misinformation represents fears and concerns about our livelihoods, health, and environment, and they need to be taken seriously.
- The basic issues (disputes) in policy dialogues center on three issues:
  - Allocation of rights to resources
  - Distribution of benefits and costs
  - Balancing of economic, social, and environmental pillars
- Typically those kinds of disputes have a series of common attributes that make them difficult to deal with in established forums for decisionmaking, particularly in nation-states, judiciaries, and legislatures:
  - Long-term horizons
  - Multiple jurisdictions (crossborder issues, including borders within the country)
  - Science intensiveness
  - Potentially large impact on vulnerable populations

• Multitakeholder processes
- Multistakeholder processes (MSPs) are designed to address the foregoing challenges.
- MSPs started gaining currency and ground at the Rio Earth Summit in that they were formally endorsed as a legitimate and necessary way to arrive at a different way of decisionmaking.
- The multistakeholder idea directly stems from negotiation theory and practice. Participants in a dialogue are not only exchanging information, but learning more and trying to achieve a joint objective. Participants can bargain over the exchange of resources, make joint decisions, and have mutual influence. There is a need for some form of interdependence. Does each individual here in the room believe he or she can achieve his or her strategies (corporate or otherwise) without other constituents in society? If so, then he or she is not interdependent and cannot fully participate in the process.
- The conventional wisdom about negotiation is that it is an adversarial relationship—what one gains, the other loses. Participants often artificially inflate demands, trade concessions grudgingly, show no empathy, and challenge the legitimacy of others' claims. However, the intuition behind multistakeholder dialogues is that there is another way—one that will make people not worse off and, one hopes, better off.
- But how are MSPs conducted? How do we know that we have a good outcome?

- A way of evaluating results is to strive for a process that is fair, efficient, wise (well-informed), and stable (so that it does not fall apart after an agreement is signed).
- Fairness includes due process, transparency in the process, predictability of the proceedings, and protection of confidentiality as much as possible.
- What is a good outcome? One answer is that participants should be at least as well off as without the process and, one hopes, better and not worse.
- Why have a policy dialogue? Because better decisions are based on efficiency, equity, wisdom, and stability. A policy dialogue has staying power and can be translated to other parts of the world in terms of the momentum of the process and the learning that was achieved during the process. It has legitimacy, and there is a certain amount of ownership on the part of the stakeholders because they feel that they were heard in the process.
- A policy dialogue constitutes one answer known as the crisis of implementation. The World Summit on Sustainable Development was supposed to be the "implementation summit," as very little of what came out of Rio was implemented. One of the answers found was that governments cannot accomplish implementation by themselves. Business, civil society, and other actors have resources and knowledge and must be brought in if implementation is to happen. It is the same intuition that is behind partnerships and the integration of other actors, which are so fashionable now at the implementation level.
- There are several key procedural suggestions for effective multistakeholder approaches. Informality and a meeting space in a nice place far away from where we usually are create a different atmosphere. The procedure should be collaborative rather than adversarial. Skilled third-party assistance is absolutely crucial in the process in order to create a good atmosphere. An emphasis should be placed on analysis, not on how participants feel. There are a lot of questions and things that are unknown—what are those questions? Are there ways to jointly frame those questions and jointly answer them? Are there common ways of approaching the problems that can be defined jointly? Protection should be provided against pressures for participants to play directly to their constituencies. Transparency does not mean that each and every statement made in a room goes out to the public and the media (then the audience is not the people in the room, but the people outside). The process should be protected until participants are at the point at which they are ready to let their constituencies/communities know what came out of the process.

- Dialogues or negotiations are divided into two phases: (1) creating value, increasing the pie, and identifying more common interests and (2) deciding whatever has been achieved.
- Several process questions need to be considered as this group moves forward.
- We are in phase one. Questions to be asked include these:
  - Is there a compelling issue that needs to be addressed?
  - Does it need to be addressed through a policy dialogue? (One answer we heard in the discussion this morning was "No." Maybe other people have different thoughts about this.)
  - If we do not do anything, what will happen?
  - Are the people here in the room and other stakeholders actually committed to continuing this process? (One of the things we heard is that we do not want a lot more meetings; we want action on the ground. However a policy dialogue is all about having meetings to try to exchange ideas and move beyond one-time, one-shot deals.)
  - Do people have the resources and the interest to move forward?
  - If yes, how do we do this? Define a purpose. (This is part of what we are trying to accomplish at this meeting.)
  - What are the dialogue's objectives, tasks, and products?
    - What are the ground rules? Who should be a part of the dialogue, and how do participants engage each other?
    - What type of institutional structure should be used?
    - How will others be drawn into this process?
    - Should we have an issues assessment or a broader consultation before we zero in on certain issues that we think are priorities?
    - Should we have a steering committee? What should be its terms of reference?
    - What are our meeting procedures?
    - What types of interaction should we have with the media?
- Then the dialogue would move into the operational phase, in which participants would clarify their interests and common understandings and recognize the need for discussion away from the table (i.e., individual meetings, Internet processes). These are the kinds of questions the group needs to think about in order to move ahead in this process and have it considered transparent and legitimate. Very often this is not what happens; very often people get invited and then they go home.
- A few process problems that often arise, and that do not have any magic answers, should be flagged:

- *Representation.* Who speaks for whom? What kind of accountability is there? What kind of standards are there for participation? Would international NGOs, such as Greenpeace, or southern African NGOs be more plausible actors in this policy dialogue? Or perhaps both? Remember to be mindful of the legitimacy concern.
- *The link to official decisionmaking.* Government has legitimate concerns about their decisionmaking power, and they are resistant to stakeholder inclusion. There are very defined rules about who gets to be part of the conversation and who does not. Multistakeholder processes throw all of this up in the air, and it is not obvious that someone who claims they speak for a particular group of people actually does. One answer is to create circles of engagement.
- *Do you want an ad hoc body or permanent body?* There is the possibility of moving the dialogue along into something a little bit more stable and institutionalized.
- *Who are the stakeholders?* It is not just individuals, groups, and organizations who have an interest in the issue at hand, or have a responsibility to make a decision on an issue. It is also—and this is very important—those who have the power to thwart a decision.
- *Knowledge integration.* Which information is considered legitimate?
- *Resource mobilization and capacity building.* Are there things we need to do to make sure other stakeholders' voices are heard? Do we have a responsibility to do these things?

Moderated Plenary Discussion

Several points raised by Ms. Ferenz were further discussed by participants, including the nature of the policy dialogue process and issues revolving around authorization, reporting, legitimacy, funding, and participation in the dialogue.

The process of the policy dialogue was described as knowledge-intensive and nonlinear. An amount of information is available, and a range of stakeholders have to manage the complexity of the issues. The process does not start at point A and end at point Z, with the same agenda throughout the process. It is full of uncertainty; the outcome is not predetermined but rather changes depending upon the interests of the stakeholders. The process was also described as a collective learning process involving self-discovery and joint problem solving. The complex political environment needs to be appreciated in terms of how the stakeholders are managed. It should be understood that consensus may not arrive at the end and that a singular outcome should not be focused upon. Instead, consensus building should be the aim, recognizing that a range of intermediate outcomes will be generated along the way.

Meetings involving pro-GMO and anti-GMO groups in Zimbabwe were used as an illustrative example of groups' being able to debate and actually inform each other. Participants noted that this dialogue must have meetings constructed to enable people to focus beyond their positions and instead look at their own interests or those of their constituencies. Meetings should be moderated with the understanding that the group is on a negotiating platform, and participants should not be afraid to voice their interests and opinions. If trust and respect are created, participants will be able to find a common line.

It was also pointed out that there are alternatives to multistakeholder dialogues. For example, (a) a science-based workshop with conclusions relayed to policymakers; (b) a parliamentary hearing with subcommittees on agriculture and health, which prepares a lawmaking initiative that is then pondered broadly and across parliaments' factions; (c) open town hall meetings with delegates; (d) electronic dialogues, which are totally open to those who have access to the Internet (which may not be ideal for this region); or (e) media briefings and working indirectly through journalism, which can be a good facilitator of the dialogue. It was suggested that these other mechanisms be kept on the table, particularly in the interest of the regional culture. It was also proposed that a smaller committee be formed, which would determine, given constraints, the best possible feasible option based on the ideal presented by Ms. Ferenz. This may be a hybrid approach, adjusted to the cultural and political situation and the context of media, parliaments, and science. However, it was also pointed out that in considering alternate options the objective of the policy dialogue must be examined. Not all of the previously outlined alternatives have the same objective. For example, if the media are used as facilitators, this brings the dialogue into the advocacy realm and not the consensus-building realm.

Also in terms of the process, the idea of a neutral moderator was questioned, and it was suggested that the interests of the organization from which that person comes should be examined. However, it was also pointed out that moderators are often chosen by the group and can be dismissed by the group at any time. Therefore, the moderator has an incentive to keep various stakeholders engaged and to facilitate a fair process, because these issues will have professional effects for him or her. Some people moderate meetings for a living. It was also pointed out that there are ways to create teams of people/process managers that keep each other balanced and honest if the group distrusts the impartiality of one organization or individual.

Another key issue raised by the group was that this policy dialogue process needs to fit into the environment and context of the region. It was pointed out that the ideal process presented may need to be altered considering different policymaking cultures. Given that in some countries there is not such a thing as participation in policymaking, will different policy environments determine different processes? It was questioned whether governments in southern Africa have decided to have a common policy or whether it is policy to not make any key decisions on this set of biotechnology issues for certain reasons. It was suggested that the group consider the different alternatives outlined previously and which would be most appropriate under what circumstances. It was also suggested that existing processes should be examined and lessons drawn from them. On the other hand, Ms. Ferenz pointed out that there is a danger of conflating culture and a certain kind of government structure. One argument is that it is not culturally acceptable to have certain types of consultations, but the idea of a policy dialogue should not be rejected on that basis. She indicated that her organization has brought the principles of multi-stakeholder dialogues to various parts of the world, including the Arab world, and it is not impossible; it just involves considering which parts of the process to adapt.

The group also considered the costs of such a process. It was acknowledged that in the short term these types of process are costly. A participant pointed out that in his country it took two years of consultation to pass one piece of legislation; it has been six months since it was passed, and it is yet to be implemented because consultations are being undertaken about how it is going to work on the ground. Is this process something that can be afforded? Costs also influence the mechanisms chosen for consultation. There are also costs of *not* undertaking the policy dialogue. What are the opportunity costs? And if a dialogue does go ahead, who pays the financial costs? How neutral is the funding agency?

Participants also raised the question of who authorizes the process. Is it a group of scientists that will essentially create a task force or panel to manage this process? Who is the client? Who is going to see the final product? Where will the group report in terms of expectations? Participants also asked who makes decisions in a consensus-building model. Who moves the common agreement or understanding reached by the group forward? In response, it was pointed out that any process requiring government action must go through government channels. However, participants in the dialogue can work in partnerships and each take a responsibility and move forward in a certain direction. It was indicated that what usually happens is that a plan is elaborated and presented to official decisionmakers, such as ministries or parliaments at the national level or the WTO at the international level. The group should identify which institutional bodies are empowered by the national community and the larger international community.

Linked to this discussion was the issue of the legitimacy of the process. A participant suggested that the group not be too shy in establishing a legitimate process, particularly given that some countries may approach this process with concern and criticism. Participants asked how "official" arguments coming out of the process would be. How would certain participants not at the table be dealt with in terms of legitimacy? The importance of the element of trust was highlighted. It was pointed out that the issue of trust has not been well researched in terms of what role it plays in the acceptance of technologies; however, some research has indicated that the acceptance of technologies and persons is based on two elements: competence and trust, the strategic optimum being right in between. If the group talks only about the benefits, but not about the risks, trust is sacrificed. It is a function of how the group communicates. It was pointed out that having a diverse, multistakeholder body come to an agreement would enhance the public's trust and perhaps would enhance claims of legitimacy and competence at the regional or national levels.

Dr. von Braun indicated that FANRPAN and IFPRI carefully considered issues of funding and legitimacy when planning the workshop, and took the position that the workshop would be funded only by IFPRI resources, although there were indications that other donors would be willing to fund. He also pointed out that a self-selected internationally composed board of trustees governs IFPRI, and their composition and governance structures are transparent and public. It was also indicated that FANRPAN has a similarly legitimate governance structure. Dr. von Braun also pointed out that Dr. Mugabe was asked to chair the session not only because he is a skilled moderator, but also because his participation and the participation of NEPAD bring an Africawide legitimacy. He suggested that once a structure has been established, the group can approach other donors and there will not be a problem with legitimacy.

A participant suggested that funds from other sources come through one pot of general funding. This is preferred over direct funding from multinational private-sector companies, for example, because if they are direct donors, certain governments may not participate.

Another important topic discussed by participants was the issue of who is invited to the table. It was pointed out that very often the largest sector of the public (consumers) are the ones who are left out. There is then a problem of translating what has been discussed back onto the ground, and this can breed mistrust. It was suggested that the process be publicized as widely as possible, such as by listserv or the media, and that information on who else to include in the process be sought. Often the argument of lack of representativeness is brought out at the end of a dialogue in order to undermine the process. To avoid this criticism, stakeholders should be sought out and opportunities provided.

The group discussed the likely unevenness in understanding about biotechnology across participants. This places greater emphasis on facilitation and tailored awareness building. It was suggested that stakeholders be brought together before sitting down at a dialogue meeting to allow them to understand the subject in a uniform manner and to create awareness. On the other hand, it was pointed out that participants would never be on equal terms at a high science level. However, it was indicated that one of the underlying assumptions about the inclusion of all stakeholders is that participants know on some level how something affects them, and although this may be a different kind of knowledge, this is where their input to the dialogue comes from. All stakeholders have concerns, questions, and fears that they can raise. It is the challenge of the moderator to bring all of the concerns onto the table and to facilitate an integrative process whereby practical issues can be on the agenda just as the policy issues are. These types of groups should not have to learn how to "speak the speak" to join.

Participants also asked whether the powerful could negotiate with the weak. The weak are the rural minority, the farmers; the powerful are the ones who have the scientific knowledge. How can that power be balanced? The weak may not be weak in terms of their opinions, but their circumstances have rendered them so. How do we structure the process to allow them to engage? It was pointed out that no process could fully get rid of power differentials. The question is whether an interaction of this sort is better for the powerless than an interaction of another sort. The mechanisms through which to achieve parity are the ground rules of the dialogue, which should ensure that equal time and equal space are given to everyone around the table, and those ground rules should be enforced by someone. A participant indicated that in his experience, differences between the farmers and the high-level policymakers did not show up in overall discussions. As long as the right environment and process are set up, everyone can come to the table and contribute. Another participant indicated that the group should be sensitive to the feelings of people it thinks are at a different level, as in her experience farmers often deeply resent the inference that they are not capable of absorbing some of the science that scientists can absorb. As many people as possible need to be brought to the table, because once the people are familiar with the technology, they will eventually appreciate it and consider whether they want it themselves.

Other points raised during the discussion revolved around what types of issues should be on the agenda and how they should be framed. It was pointed out that there are four different levels of decisionmaking that the group may want to address: (a) whether to invest in or permit biotechnology in agricultural systems, (b) how to regulate it, (c) what traits to develop in biotechnology research (which need to be grounded in the reality of the field), and (d) how to facilitate adoption of the technology by the end user (which also needs to be grounded in the reality of the field). It was suggested that the group consider an element of investment in order to make the approach different than those used for other initiatives. There is a case to be made that we should consider investing in skills that are important for the group to have in order to move forward and make a difference. Finally, the question of timing was raised. How easy or difficult would it be to reach consensus given the issue and timing? It was suggested that a central notion of the multistakeholder approach might be to eliminate the time dimension.

Information Sharing: National and Regional Experiences

Dr. Mugabe indicated that through the group's discussions it had become clear that there are a variety of processes at the national and the subregional levels and that countries are experimenting. He therefore proposed that the group spend time to share experiences to learn why some countries decided to form particular kinds of groups and what kinds of policy issues those processes are addressing.

Namibia

Dr. Martha Kandawa-Schulz relayed the experience of Namibia. She indicated that when Namibia got funding from UNEP to develop a policy framework, they decided that they did not know enough about what was happening in their country itself. So they developed a country study first and decided to then develop the policy. Upon the study's completion, they started working on the national policy. There was an eight-month debate about whether to call it a biotechnology policy or a biosafety policy, and in the end it was termed a policy "enabling the safe use of biotechnology." The cabinet passed the policy in 1999. After that, there was discussion about which ministry was the competent authority for biosafety issues. It was decided that the Ministry of Science and Technology would be responsible for administering the law. A meeting was planned two weeks after this meeting regarding Namibia's biotechnology strategy and what it should include in terms of content. Based on the points discussed at the meeting, a biotechnology strategy will be drafted and will be linked with the biosafety policy. The public was involved through the use of the biodiversity program, which joins many stakeholders from 13 groups. Following that, workshops were held with smaller groups starting with farmers, scientists, and so on, and then a big workshop was held at the end at which all the stakeholders were together. Groups such as farmers' unions, the meat board, and the agronomic board were brought together separately before being brought together with the larger group, and the topic and goals of the dialogue were conveyed to them before the groups were joined together. The commission includes representatives of the Biotechnology Alliance (which is one of the working groups), government institutions (ministries of the environment, agriculture, fisheries, trade, and health), the private sector, parastatals, the university, and the consumer lobby (which has been active for the last three years).

Kenya

Prof. Norah Olembo spoke about the experience of the biotechnology policy development process in Kenya. She indicated that it has been about 10 years since the process was initiated, when they first heard a lot about biotechnology and its elements and decided that they wanted to use the technology. It soon dawned on them that they needed to have biosafety regulations in place to guide them; therefore, they started the process of forming a policy. The Netherlands gave them money to form a committee of experts, which had a lot of meetings with farmers, scientists, and industrial organizations for an entire year to see what those stakeholders thought of the technology before they embarked upon it. At this point in time, all biotechnology was being considered, not only GMOs. The people asked about the risks and the benefits and decided they wanted to use it. The committee looked into it and gathered literature from various organizations that articulated guiding regulations. The World Bank donated the documents. This material was used as a baseline to develop biosafety rules, which were drawn up under the National Council of Science and Technology as supervised by the Ministry of Science and Technology. The rules were stringent and a bit restrictive, but they worked and have worked ever since. A national committee that is recognized at the government level deals with these applications. There are guidelines as to what is supposed to be asked and questions that come to the committee.

Kenya has now reached the level of field experimentation, and there is insistence that there be a laboratory-experimental stage to test technologies before they go out into the field. A specific laboratory is being built for this purpose. There is also dummy field experimentation, a trial without the GMOs themselves, for the sake of the farmers who want to know whether the technology is safe. Groups have been taken to the field trials from surrounding communities to familiarize themselves with the work, to ask questions, and to better understand how safe it is for them. These types of activities have been ongoing, and one project for insect-resistant maize is already underway in Kenya. The committee has also received an application for *Bt* cotton, which is moving faster. Modifications with carnations and a few other projects are in the pipeline. A new sweet potato came through the system and is still at the lab level, but not yet with the farmers. The trial stage is advanced, but the product is not yet in the field.

Prof. Olembo concluded by noting that the committee felt it was very useful to involve as many stakeholders as possible for acceptance of the technology. She noted that it is also useful to be serious in dealing with an idea and carry it to the end. The committee acknowledged that there were dangers to expect and that they had to put structures in place to address them. She also mentioned a publicity and education exercise in conjunction with the African Biotechnology Stakeholders

Forum, which is involved with the news media and other stakeholders, including parliamentarians.

When asked about the process used to approve applications to conduct GMO trials, Prof. Olembo noted that there are conditions and requirements in the biosafety guidelines that applications have to satisfy before they can move into the trial stage. She noted that the *Bt* maize took three years to go through the process, and that the first application does not necessarily qualify. The committee is in the process of revising the regulations using what they have learned over the past six years of use. Prof. Olembo also indicated that there is one set of guidelines for all activities. Conditions for approval of the application include such things as whether it has been accepted in the country of origin, evidence of risk assessment and experimentation, and proof of where the technology has come from. The guidelines will be different when the technology is created within the country. The committee insists that local institutions have their own biosafety guidelines that feed into the national guidelines. For instance, KARI has a set of biosafety regulations that deal with all the nitty-gritty requirements for food safety, such as how to dispose of materials.

The question of whether the regulations consider consumer health was raised. For example, particularly for *Bt* corn, the country from which the technology comes may not necessarily be a corn-eating country, whereas the country accepting the technology may eat a lot of corn and may not be happy with those regulations. It was asked whether Kenya is undertaking any human or animal safety studies, and it was noted that most of the animal studies done in Germany show no traces of *Bt* in the protein of the animals. Prof. Olembo noted that applications from outside are required to provide evidence of all tests undertaken in the country of origin, but Kenya does not carry out any tests. They have not reached the stage of carrying out experimentation on animals in Kenya itself. Perhaps this will be introduced at the stage of laboratory testing so that researchers can consider the effects that GMOs have on Kenyan animals in lab conditions.

A participant raised the point that although products are now considered safe for the countries of this region if they are considered safe in the country of origin, testing should be based on the particular characteristics of a product rather than the process by which it was created, and products should be evaluated in the context in which they will be used. In this region, that context includes a high rate of morbidity, which affects the absorption of toxins; a high prevalence of HIV, which involves the immune system; and a high rate of malnutrition. The U.S. Food and Drug Administration policy focuses on the U.S. population and is blind to the conditions, diets, and food habits elsewhere that *Bt* products might be used. Because many countries in the region base 50 to 70 percent of their diet on one product, the population using the product may be totally exposed, and because the product is not extensively used in the country of origin, the risk assessments undertaken in the country of origin may not be sufficient.

Participants discussed the need to make people aware of the implications of adopting biotechnology, which brought them back to the idea of stakeholder involvement. It was noted that if people are serious about food security in Africa, chances must be taken, but in an informed matter. Kenya has undertaken cautious steps over a long period of time because the process needs to begin somewhere if the people are to be convinced that they need the technology and are serious about change.

When asked about who enforces the regulations in Kenya, Prof. Olembo indicated that there is an interdisciplinary committee comprised of many different stakeholders. When an application is submitted, there is initial work to see that all the required papers are there, and if necessary the applicants are asked for more information. Once completed, the application is presented to the committee, which scrutinizes it very closely. The committee has received complaints that the process is taking too long, but it wants to be sure that Kenyans are safe and ready to move on to the next stage.

A participant asked whether Kenya is monitoring at entry points and whether monitoring has been built into the law and, if so, what the institutional framework is. Prof. Olembo acknowledged that monitoring is a difficult thing to do. She noted that people in Kenya are not too worried about seeds coming in through someone's pocket and also that it is very difficult to monitor seeds coming in by way of donations. Should there be testing kits at the entry points to determine whether a product is GM? If any maize or soybean product has been imported from the United States or elsewhere, it will most likely have GM content. It is not easy to say a country is completely free of GMOs. Prof. Olembo indicated that Kenya does not have the capacity for testing incoming foodstuffs.

Participants also discussed the issue of labeling. Prof. Olembo noted that Kenya recently ratified the Cartagena Protocol, which does have guidelines for labeling. This subject will be debated at the national level to see whether it is compulsory to adhere to the guidelines of the protocol. At the moment, food is not labeled on the shelves in Kenya (whether GM or non-GM). Dr. Schulz indicated that Namibia's draft law says that food has to be clearly labeled so at least consumers can see whether it is GM. The law states that if it is a normal GM product, it must go to the registrar, and the ministries are modifying their regulation forms so they can include a GM indication.

Dr. Schulz also pointed out that countries will have to follow labeling procedures at the SADC level and suggested that labeling be coordinated regionally, as a lot of goods are traded within the region. A participant raised the issue of how decisions are made and how trade-offs are evaluated in terms of trade issues, considering for example whether a country wants to undertake a GMO trial on a good that might contaminate exports to the European Union. Prof. Olembo acknowledged that in Kenya, when it comes to a product that is obviously intended for export, those considerations would have to be taken into account. But with maize it was found that there would be no way that Kenya would be an exporter of maize in the near future. However, she noted that the issue might be more relevant to other products.

SADC Advisory Committee on Biotechnology

Dr. Bernard Luhanga presented information on the creation of the SADC Advisory Committee on Biotechnology. He noted that biotechnology issues were first put on the agenda of the Council of Ministers due to the humanitarian crisis over the past three years. In August 2001 they recognized that there would be a production shortfall within the region, particularly affecting Malawi, Zambia, and Zimbabwe, and that this was due mostly to drought and simultaneous flooding, particularly in Malawi, and the lack of regional surpluses at the time. A directive was issued asking that the immediate problem of the humanitarian crisis be addressed, but also that a long-term strategy to deal with the food insecurity situation in the region be developed. When the next food security report was presented in 2002–03, the situation had further deteriorated. Six countries were now affected, and, in terms of human cost, 40 million people were at risk.

A directive called for the ministers of agriculture to meet and come up with a strategy, so there was a special meeting to figure out what to do. Since the region had no surpluses, imports had to come from the outside, and the major donor was the United States, which was obviously giving GM food aid (or maybe it was not, but the food was not labeled one way or another; the expectation in the region was that it was GM), the question arose as to how the SADC should handle the issue of GMOs. Each country has a sovereign right to determine whether to accept GMOs, but if there are transit arrangements, it needs to be discussed with neighbors. Zambia took a stand on GMOs, and other countries' stands became very clear. The ministers wanted to put in place mechanisms of accepting GMOs under certain conditions, and a decision was deliberately taken to look at the need for and potential promise and risks of GMOs. They recommended to the council that an advisory committee on biotechnology be formed.

Dr. Unesu Ushewokunze-Obatolu provided further details about the committee. She indicated that their first meeting had been held the previous week, and they have yet to produce an official record of the proceedings. The advisory committee was nominated in February; it has a membership of eight and is serviced by the SADC Secretariat, through which it reports to the Council of Ministers. Its mandate is to advise the SADC on all issues having to do with biotechnology and biosafety. Given information from the secretary general of the SADC, the committee's authority enables them to inquire from each country about its progress or any assistance the countries may need from time to time, and also to seek advice about where the committee might get professional advice. Although the committee's members are from different backgrounds and quite diverse, they cannot cater to all relevant areas so they may need to seek outside advice. One committee member is a lawyer who has been exposed to trade/biotechnology/IPR issues.

The terms of reference of the committee have been drafted for their review. It was decided that in order to come up with the best advice, they would look at policies, legislation, and regulations in view of the fact that each country of the SADC region should have legislation in place on biosafety in order to receive or regulate activities on biotechnology by the year 2004. The committee also agreed to look closely at the Africa Union model law that was drafted some years ago and to examine how best to integrate the requirements of the Cartagena Protocol on Biosafety. The committee will also look at ways of institutionalizing processes that have to do with activities that affect biosafety within the region, encouraging each country to set institutional mechanisms in place. They will also look at the resources available, because that will drive the process. These tasks are additional responsibilities in areas that are already regulated by a number of sectors in each of the different countries. The committee will also look at the information resources that can be used in advising or can be accessed by the various countries at the regional level. They will consider human capacity, and particularly capacity building. Expertise levels are expected to be low, although it is unsure at the moment, before the literature is reviewed. The key question the committee will examine is how to put biotechnology to good use, realizing the comparative advantages between and among the countries and looking at different ways of mobilizing financial resources.

A number of organizations are interested in looking at different aspects of biotechnology, provided the committee can sit down and learn what the priorities are. It was decided to make a strong recommendation that the region itself set out to commit resources before looking elsewhere. The committee also considered the issue of public awareness, particularly that of the region, and agreed it is the small farmer who most needs awareness at this time (this is not to say that it is important only to small farmers; urban areas are also affected).

The committee will maintain oversight of the progress each of the countries will be making over time in developing systems to implement biosafety, in particular in areas such as knowledge development, research, and capacity building. They have set out to engage in various activities, but have not yet concluded a full plan of action. A meeting will be held in June, but before then it was agreed that they would undertake a number of reviews so the committee will have a baseline set of information about the region in terms of where the region is. And they will take stock of the inventories relating to the various resources already available and identify the gaps that exist so that they can make appropriate recommendations to the sectors. A specific assignment is to look at issues that will impact the design of a SADC regional model. Having been closely associated with the Africa Union (AU) area, a number of members are looking at various issues, such as environmental impact, public health impact, food safety, and consumer concerns. When the committee meets again, they will decide which of these issues can translate into policy instruments.

Participants expressed concern over the 2004 deadline for regulations. In response to a question about whether the committee will do anything to help countries put that administration in place, Dr. Luhanga responded that this issue was of high priority in the action plan during the previous meeting, and members are developing some concepts as to how to approach the issue. One approach is to have each country come up with its own legislation (with input from the legislation of other member states). The options are being put into concept form, and it is hoped that they will be discussed in June or July, at which time member states would be free to choose one. Dr. Luhanga noted that these are national decisions, but definitely the deadline is there. Another participant did not see how the deadline could be met, especially given the lengthy process for advancing draft forms of legislation. It was also asked whether the Council of Ministers has informed countries of ways of acquiring funding to undertake the development of the legislation.

It was pointed out that those involved are confident that funds will be found somewhere, and full information is not available yet. Efforts are already underway in four of the countries who are using UNEP funding. Other donors have expressed interest, but the committee must sit down and decide for exactly what the funds will be used.

The question was raised about what forms of support (not only funding) research institutions like IFPRI could offer. Dr. Ushewokunze-Obatolu noted that there is a lot of experience and expertise on policy development and international debates within IFPRI and other organizations such as the International Service for National Agricultural Research, and the committee is investigating where they can tap into it in certain areas. Dr. Luhanga added that information management is one area in which expertise may be needed. He noted that the committee is now looking at institutional frameworks, and this assessment, which is very important for the region, is one to which IFPRI may be able to add value.

Concern was raised over certain elements of the AU model law, which would make it difficult for companies to invest in biotechnology because it is totally unattractive to industry. A participant asked if there was room on the committee to involve other stakeholders such as industry. Dr. Luhanga indicated that the AU model is only a reference, not the key document, and that the committee is going to ask for input from all stakeholders. They will be coming up with their own model that reflects what is happening on the ground, which each country can domesticate.

Participants expressed support for the SADC initiative, indicating that the issue of harmonization is key. It was also pointed out that the group should ensure that the dialogue process beginning with this meeting should complement the national and regional processes already underway.

Day 1 Closing Remarks

The Chair asked Dr. von Braun to make some closing remarks at the end of the first day. Dr. von Braun noted that the process started is extremely useful and should continue as a quasi-independent, not mainly government-driven, process of dialogue. He also suggested that the group stick to the term *dialogue* and not use *negotiation*, as he felt the word *negotiation* was a bit too heavy and too loaded.

Dr. von Braun also suggested that this process continue as one that is driven by a set of international and regional organizations that are partially independent of government-driven processes. He proposed that FANRPAN, IFPRI, and NEPAD be umbrella organizations for this process, but said that every voice at the table should count. The process should be one that remains as open and broad-ranging as it is today, and should probably be even broader in terms of participants. Consumer groups, farmer groups, and people engaged in trade and food industries who are part of the decisionmaking, agenda-setting communities should be added in future meetings.

Dr. von Braun proposed that some sort of a working group or committee with which the group around the table would be comfortable grow out of the dialogue. The committee could continue the work between meetings and would have no more than five or seven people. He suggested that by the conclusion of the second day of the meeting the group create such a working committee to move the process forward, synthesize the agenda of the dialogue, and aggregate the conclusions of this meeting. The committee could take on initiatives such as e-mail dialogues and liaise with other groups engaged in activities of the same nature in the region.

Dr. von Braun also commended the Chair, Dr. Mugabe, and suggested that he be asked to chair this committee.

Dr. von Braun also noted that the group would have to explain to the rest of the world why it is undertaking this dialogue, and perhaps that should be further reflected upon in the coming discussions. Although there are stakeholder dialogues driven by governments, international finance, UN organizations (with intergovernmental characteristics), or foundations (which are more or less close to industry groups), what is missing from this mix of dialogues is exactly the type of exercise conducted at this meeting. Dr. von Braun suggested that the exercise be carried on, with a sunset clause to the effect that it should reassess itself at the end of 2004. He noted that the committee should be entrusted with identifying a few milestones, such as when to deliver what, so that progress can be measured. Dr. von Braun concluded by noting that his suggestions are preliminary and should be further discussed.

The Chair closed the day's proceedings, thanking the speakers for their presentations and participants for their discussion.

DAY 2

Introduction to Day 2 (Dr. John Mugabe, Moderator)

Dr. Mugabe opened Day 2 by reminding participants of the issues that were raised during the first day's discussions. He raised the following topics:

- The appropriate subjects of debate. In terms of policy issues, discussions were guided by the first presentation, which indicated that a number of African countries are in fact embracing biotechnology, although they are currently at different stages of biotechnology. The debate should not be on whether these countries should be investing in biotechnology, but on how these countries can maximize benefits and minimize risks through the development of biotechnology.

- The formulation of biosafety policies and frameworks regarding the use of these technologies. Drafts and legislation are currently being developed.

- The need to harmonize policies, given transboundary and trade issues.

- Trade liberalization and its implications vis-à-vis the regulation of GM food. In terms of information and experiences, this is one area in which more research and analysis are needed. Many of the countries in the region may be confronted with conflicts between their efforts to liberalize their economies and those to develop and use biotechnology. In broad terms, the WTO and biosafety frameworks need to be examined to find out how the two can evolve at the domestic level.

- Expansion of the knowledge base regarding the implications of research and development on biotechnology and distribution of the benefits of the technology.

- The sharing of best practices for assessment on the national level.

- Issues involving the costs and benefits (particularly economic) of biotechnology. Although they were not discussed, these issues were flagged in the background papers, and they should be conveyed to policymakers.

- The importance of ensuring that the introduction of biotechnology does not in any way undermine local, indigenous technology.

- The need to build a platform or platforms for dialogue and, where possible, consensus building, on the range of unresolved issues. Such platform(s) may facilitate interactions between ongoing national and subregional efforts. From the discussions it was clear that at the national and the regional levels there are some policy processes that may generate consensus, but most of these processes are those of governmental committees, with an emphasis on risk assessment. In those processes less emphasis is placed on some of the issues discussed in this dialogue, such as intellectual property regimes.

- Inclusion in the debate of those groups that are not participating.

- The establishment by an institution such as IFPRI, together with others, of a small committee that will be tasked with guiding the regional dialogue. The process may not necessarily aim at consensus building but rather at awareness, interaction, and exchange of information, and perhaps at influencing particular policies.

Dr. Mugabe then asked participants to consider several questions as the dialogue process moves forward:

- What would the mandate or role of a committee be? The general role would be to facilitate the dialogue process, but what would the specific roles be?

- What issues should be high on the agenda of the committee for discussion at the regional policy dialogues?

- How would the committee be composed? The first impression is that IFPRI and others would be conveners of this committee, but the members would be from this group and also drawn from others. How would we accept and guide the membership of the committee?

- How would this dialogue relate to other stakeholder processes?

- To whom would the committee report? Would it report to the broader constituents? Would all of the minutes/reports be accessible to all stakeholders?

- How long should the committee be in place? Should it be open-ended?

- Are there any key policy arenas that this dialogue or process should seek to influence, at least in the short term? For instance, those of NEPAD, the SADC, the WTO, the FAO? Where do we find the policy champions? The dialogue needs to identify policy arenas so that it can make an impact on policy, and this must be kept in mind in terms of the processes the group wants to influence in the short and medium terms.

Priority Policy Issues

The group discussed and agreed upon the following list of priority policy issues that could potentially be explored by the policy dialogue. The dimensions of each issue were then considered and adapted to form the two tables that follow (Tables A.1 and A.2). However, it should be kept in mind that although it may be helpful to frame the issues in the two tables, not all of the topics are included in the tables.

Table A.1 Emerging priority policy issues

	Clustered policy issues			
Clustered activities	Biosafety	Intellectual property protection	Trade	Technology development and transfer
Information gathering, exchange, and analysis				
Capacity building/infrastructure				
Harmonization				
Cooperation/collaboration				

260 APPENDIX A

Table A.2 Biotechnology development for food security

Food security needs addressed by biotechnology	Technology development strategies		
	Research	Technological development	Technology transfer/diffusion
Drought			
Soil fertility			
Malnutrition			
Pests and diseases			

The priority issues identified as follows will be used as a provisional list for the committee to consider for future dialogues.

- Biosafety policies and frameworks

- Harmonization

- Trade issues

- Intellectual property rights

- Risk assessment

- Economic costs and benefits

- Local technology

- Links to national/regional development strategies

- Biotechnology and food security

- Development of biotechnology strategy (proactive vs. reactive)

- Seed (access, availability, policies, trade)

- Access to germ plasm

- Liability and redress (public and private)

- Protection and conservation of biodiversity

- Public- and private-sector roles

- The policy formulation process

   Various points were raised during the development of the list of priority policy issues, including these:

- The way in which the policy dialogue relates to and feeds into existing processes, such as those of the SADC Advisory Committee on Biotechnology, should be kept in mind. The dialogue should feed into that committee and follow up on the progress of the committee.

- The issues of seed industries and seed production should be considered. The Rockefeller Foundation has funding in technology transfer and is trying to facilitate the development of patenting rights to move some materials that are not necessarily commercially viable for a commercial seed company to produce.

- Issues of intellectual property rights include how to ensure that technologies are protected and made available to those that require them and also how to provide information on the range of technology through the use of domestic databases.

- That the dialogue should provide feedback to research organizations in the region and internationally should be one of its core principles. There are large knowledge/research gaps related to biosafety; for instance, (1) there is practically no understanding of the relationships between *Bt* GM crops and soils (the basic research has not been done, though soil scientists say it is very important) and (2) there has been practically no basic research on the whole range of food safety concerns that are laid out in the background paper of David Pelletier (these concerns relate to the use of GMOs under the African conditions of the vulnerable health status of populations and the very large shares of their diets from single commodities, such as corn).

- One of the most pressing issues of the dialogue should be biotechnology vis-à-vis food security needs in southern Africa. What is the contribution from all this investment in policy and regulation, and is it really addressing food security? Cotton is not going to address food security needs. Investments need to be considered in the context of national agricultural development plans.

- Promoting the harmonization of trade policy, food safety, capacity building, strategies, and so forth should be considered a core objective of the dialogue.

- It is important that the region engage in a dialogue about the incentives of the region to create its own capacities for biotechnology. Should there mainly be dialogues on biotechnologies that are reactionary (as the rest of the world invests in biotechnology and southern Africa picks and chooses)? Or should there be a policy strategy that puts the subregion itself in the driver's seat in formulating biotechnologies for subsistence farmers, which relate to agro-ecologies and drought problems, or biotechnologies for consumer health or for HIV/AIDS-burdened areas with certain micronutrient deficiencies? Perhaps both reactionary and active policies should be considered.

- The dialogue should deal not only with science issues, but also with health and safety within the national and regional strategies. A number of subregional and regional strategies have been very exclusive, and the goal should be to create a dialogue that would allow as many stakeholders as possible to feed into the process of developing a strategy. The dialogue could feed into the identification of the key target areas for influencing national and regional activities, such as the NEPAD suggestion of an African biotechnology strategy.

- These questions need to be addressed: What policy arenas should be targeted, and what is the timeline? When will African governments be making certain decisions, and how can the process of dialogue be benchmarked?

- The two key issues that require urgent action by policymakers are trade and intellectual property rights. Two other key issues are the development of biotechnology products in Africa for smallholder farmers and the development of biotechnology products for vulnerable consumers. These issues require research, investment, and capacity building. A sense of the urgency of the policy priority-setting scheme should be introduced. Because of the state of development of such technologies and the need to understand the human and biosafety issues, there is a different time dimension.

- Whether the dialogue itself should engage in or simply exchange information on public awareness activities and acceptance was widely debated. It was noted that singling out public awareness puts the policy dialogue in an advocacy role and undermines its credibility. On the other hand, it was suggested that the dialogue could exchange information on and scrutinize what countries are doing in terms of public awareness activities.

- Information and best practices should be shared on public mobilization and participation.

- The process of policy formulation should be studied. The University of Sussex Institute for Development Studies has been examining biotechnology policy processes in two countries in Africa (Kenya and Zimbabwe), and it has been suggested that another study look at other recent cases, particularly the decision-making process involved in bringing about a particular GMO policy. The dialogue can look at these kinds of studies and can synthesize information and draw lessons from them.

- Given that there were 52 meetings on biotechnology in Africa last year and a lot of information gathering is already being undertaken, the value added by the dialogue could be analysis.

- Regarding the seed issue, a parallel exercise is being undertaken by the International Maize and Wheat Improvement Center, ICRISAT, and the SADC Seed Network, which are looking at all of the seed issues, including biotechnology aspects.

It was acknowledged that the master list of priority policy issues would be considered a living document that the group could continue to put together.

Biosafety Issues

The group focused on one line item of the list of priority policy issues, that of biosafety policy and frameworks, as an example of the types of specific issues that might be addressed by dialogues on this topic. The group developed the following list of issues related to biosafety:

- Efforts to promote sharing of information and experiences, including capacity building

- The issues not yet covered, including consumer rights and safety

- Building bridges at national and subregional levels (in the areas of trade, health, environment, agriculture)

- The need for harmonization regarding trade issues

- Providing feedback to research organizations in the region and internationally

The urgency of the development of a biosafety policy was discussed. A participant noted that it is important to have biosafety guidelines, because it is advantageous to be able to use them at any time. However, it was also noted that there is a dilemma in that trade policies enforce urgency. If a country wants to trade, it has to adopt a policy immediately about whether it will accept GMOs and under what conditions. If a country is concerned about production safety and environmental safety, it must be aware that ecologies are as complex as economies, and the research agenda is so huge that it could take decades. Biosafety regulations currently focus on environmental safety; however, consumer benefits, safety, and well-being need to be further examined. A sense of urgency also arises when one considers the potential of biotechnologies to offer opportunities for improving nutrition. The current biotechnology generation focuses on the content of the product and no longer on the production characteristics only. If there are not proper biosafety and human health policies on the table, countries have no incentive to develop the technologies. Southern Africa needs to be able to trade in food and agricultural commodities inside and outside the region and therefore needs to urgently implement biosafety policies. However, there also needs to be an increased concern for people's health and food security, which also requires a sense of urgency.

It was also suggested that the process of developing biosafety policies be built into national and regional development frameworks and located within NEPAD. The dialogue should engage those other than scientists and should link with poverty eradication strategies.

A participant relayed the experience of Tanzania, which has an advanced biosafety draft document that they are hesitant to use even on an interim basis. Multinational companies using GM tobacco and other companies are putting pressure on Tanzania to accept GM products. Because regulations are not yet in place, Tanzania is not interested in accepting GM seeds. The participant stressed the need to consider facilitating the movement of GM seeds or foods. Do we want to deliver seeds or foodstuffs to people who are starving? Foodstuffs would be relevant, but seeds may be more dangerous and should be further examined. He asked why, considering the prevalence of hunger, Tanzania should not import these foodstuffs instead of letting people die. However, the participant closed by noting that we are unsure whether these foodstuffs are really good for human health.

The Chair closed the session on biosafety by noting that there was consensus on the biosafety issues that need to be taken on by the policy dialogue platform, among them information sharing, best practices, food aid, consumer rights and safety, and trade. The committee should be given the mandate to think about what other issues might be addressed under the aegis of biosafety.

Committee Mandate, Role, and Composition

Dr. Mugabe opened this session by asking the participants to consider the suggestion that a small committee be established to facilitate the regional policy dialogue and lead the process that will enable countries to ultimately develop a strategy on biotechnology for food security. It is envisioned that the committee would build bridges and engage the subregional platforms in dialogue.

Regional Scope

Participants considered the regional scope of the dialogue. Dr. von Braun asked whether it might be useful to go beyond a focus on southern Africa to a sub-Saharan or all-African perspective. In response, participants agreed that it is better to start small as a subregional exercise, with a focus on SADC countries, and then revisit. It was acknowledged that there is a trade-off between scope and depth, and it was suggested that extra resource persons from other subregions be brought into the dialogues for exchange of valuable experiences. It was noted that it would be helpful to expand future dialogues in order to feed into the NEPAD strategy.

Links with Other Initiatives

In response to a question about how the dialogue would be seen by NEPAD, Dr. Mugabe noted that NEPAD works with the subregional economic groups, so there is flexibility based on needs, although what has been stressed thus far is an Africa-wide forum for biotechnology. FARA has approached NEPAD, and discussions are taking place about whether there is scope for a subregional foundation to have a regionwide discussion on biotechnology.

It was also suggested that the committee facilitate linkages with other ongoing activities, such as those of the SADC Advisory Committee on Biotechnology, and that it make informed judgments about the dynamics of these processes and see how a dialogue of this nature could feed into other processes. Participants agreed on an active marketing strategy for the services the dialogue could provide. In other words, members of the committee would not necessarily be held captive to speak on behalf of the group, but they could enter into relationships with other stakeholders and indicate when it would be helpful for an issue to be entered into the next phase of the dialogue. Committee members could disseminate information about the potentials of the dialogue process, which can play the role of overcoming gridlocks in government debates and in debates between nongovernmental and industry circles.

Dialogue vs. Advisory Role

Participants debated the question of whether the role of the dialogue and committee is only to engage in dialogue or also to give advice on policy formulation. It was agreed that the committee is not an advisory committee; it is not making policy but making a link between the policymaking process and multistakeholder dialogue to inform particular policies. The dialogue was considered one step before an advisory body, where conflict is unresolved and where the process of dialogue can make a key contribution. The dialogues may identify recommendations and priority areas, and the committee should be seen as a supplier of that information to the decisionmakers.

Reporting/Coordination
It was agreed that the committee is accountable to the stakeholders. The reports of dialogues and syntheses generated by the committee would be distributed to all stakeholders. In a technical sense, the committee would be accountable to the three core sponsoring organizations—FANRPAN, IFPRI, and NEPAD—but these three organizations would not exercise any censorship of the outcomes of committee deliberations, nor would there be an approval process. The committee's mandate would include complete freedom to dialogue and liaise with other organizations.

Committee Mandate
Participants agreed that the committee would prepare the next dialogues, which would include (a) reviewing the initial proposal in which FANRPAN and IFPRI suggested beginning with a sequence of three dialogues, (b) determining which issues would receive priority and what aspects of those issues should be discussed in the dialogues (using the list of the priority policy issues identified by the group as well as the two tables as a framework), (c) considering whether to restrict a particular session to one stakeholder group only (i.e., parliamentarians) or whether the session should be open, (d) facilitating the commissioning of working papers on key issues around the dialogues, and (e) considering the range of key policy initiatives into which the dialogues should feed and developing a time frame.

It was also agreed that the committee would synthesize and disseminate results from dialogues, liaise with other stakeholders in other policy decisionmaking forums, and review and clarify the draft mandate, which would include the following charges:

- To maintain a regional scope, starting subregionally

- To prepare the next dialogues

- To facilitate linkages with other ongoing activities

- To synthesize and disseminate the results of dialogues

The name of the committee was left undecided for further consideration.

Nomination Criteria
Participants agreed that the committee would elect its own chair, despite the earlier proposal that Dr. Mugabe chair the committee. There was debate about whether the committee itself should have a multistakeholder membership, and it was agreed that while the committee should reflect the multistakeholder outlook of the dialogue, members would not be serving on the committee as representatives of stakeholder groups so not all groups need be represented. It was suggested that committee members in principle promote and protect the objectives of the multistakeholder dialogue (rather than their own personal viewpoints). It was agreed that individuals sit on the committee in their personal capacities.

So the committee will have the appropriate expertise or other qualities, it was also suggested that the committee members include (a) people with networking capability in the region, (b) people with experience in policy issues in the region, (c) people who are in touch with farmers and NGO groups, and (d) scientists from universities or other areas. On the other hand, it was also stated that the selection of the committee should not be restricted to such criteria, but focus more on whether the committee members can work well together and carry the process forward.

It was agreed that there would be three ex-officio members of the committee (representing the three umbrella organizations—FANRPAN, IFPRI, and NEPAD), one representative from the SADC Advisory Committee on Biotechnology, and five other members, for a maximum of nine persons presiding on the committee. It was also agreed that at least two of the five non-ex-officio members should be women.

Timeline/Benchmarks
The Chair asked participants to consider any key policy processes or key events into which the dialogue should feed. It was noted that IFPRI is planning a major conference in April 2004 on food and nutrition security in Africa, to which President Museveni would be invited. It was suggested that there might be an opportunity to link this conference with another round of the stakeholder dialogue on biotechnology. The August meeting of the ministers of agriculture was mentioned, and the group recommended preparing some informational materials to inform them about the multistakeholder dialogue initiative.

In terms of milestones, it was recommended that the committee prepare at least two more successful multistakeholder dialogues to influence the priority-setting and decisionmaking processes, one to be held in the next 6 months and the next in the 12 months thereafter. If that is not done successfully, the process can be gracefully closed in an e-mail consultation. It was noted that it will not be a failure if the institution is closed. It may even be closed if it is very successful over the next two years. The objective is not longevity, but an intensive, effective, and means-tested process, which will also be much more convincing to any potential donors. It was agreed that there should not be any predetermined outcomes for the committee, but an emphasis on making an impact where there are opportunities.

It was also suggested that the committee be entrusted to establish self-evaluation criteria and milestones. So that the committee will not serve in isolation, it was proposed that the committee set up an e-mail platform so that e-mails can be sent into a receiving pool for the committee and they can selectively answer and respond in an easy way.

Another suggestion was for the committee to oversee a preassessment of the dialogue participants' views of biotechnology, which can be revisited two years later to see if the dialogue had an impact on their views. The information could also be used as a baseline for the next dialogue to show where the group stands on certain issues. It was recommended that Dr. David Pelletier, the participant who made the suggestion, develop a questionnaire of three to five questions to which the group could respond. It was agreed that the next dialogue would have a self-assessment mechanism.

Closing Remarks

The Chair invited Dr. Takavarasha and Dr. von Braun to make a few closing remarks. The following paragraphs record what they said.

Dr. Tobias Takavarasha, FANRPAN

I would like to thank the moderator, and am grateful for the partnership between IFPRI and FANRPAN as part of a process of contributing to dialogue, debate, and advice on key policy issues. We hope to be able to forward the contributions and advice of this dialogue directly to the SADC committee or other key stakeholders.

FANRPAN is happy to give this support. The network is going through a consolidation process to continue to be well positioned to give the kind of assistance that is needed in the region—simply bridging the gap. The potential and the need for policy advice in the region are very clear to everyone, and there is a need for institutional resources, human resource support, and capacity for policy analysis.

It is hoped that the working papers prepared for this dialogue will go through a peer review process and will be published in some format. We also plan to have a short synthesis that will be circulated in the regular policy briefs of FANRPAN.

FANRPAN is also hoping to convene or be part of a meeting of permanent secretaries in the region so they can talk about key issues in the region. With that in mind, FANRPAN will continue to contact the participants in this dialogue and work with IFPRI on how to build the capacity of FANRPAN to continue to support the activities that we have undertaken.

Prof. Joachim von Braun, IFPRI
This dialogue has exceeded my expectations. I see much more clearly the potentials of multistakeholder dialogues after this experience. There were excellent dynamics over the last two days. Information was exchanged on the ongoing activities in the SADC region on biotechnology strategies and the complex issues involved in formulating and implementing biosafety policies. This meeting has made a contribution to making complex political processes better informed. I also learned a lot for other regions in the world, and at some point, maybe in two years, if this process is successful, we should compare notes on how these types of dialogues function in different parts of the world and in different cultures.

IFPRI is delighted to begin this work with FANRPAN and to do so only half a year after having signed a joint memorandum of agreement. We are equally delighted that we have expanded this to a trilateral institutional relationship between FANRPAN, IFPRI, and NEPAD on biotechnology dialogues.

IFPRI positioned itself as a facilitator, bringing knowledge from other parts of the world. I acknowledge gratefully the willingness of participants to engage at this table, due to the leadership of Dr. Mugabe but also the willingness of participants. The debates that continued over coffee and lunch breaks were a clear sign of the strong demand and need for these dialogues. The willingness of members to serve on the committee was also a strong sign of participants' willingness to engage.

This meeting will be properly documented, and I thank those who contributed behind the scenes. I would also like to highlight and specially commend Were Omamo, who cannot be with us here, as he has worked together with a team at IFPRI and FANRPAN since January to make this workshop happen.

I thank Dr. Takavarasha and Dr. Mugabe for their leadership. The meeting is formally closed.

Appendix B

Workshop Program and Steering Committee Meeting Notes

Workshop Program
Meeting location: Senators Hotel, Caesars Gauteng, Johannesburg, South Africa
Meeting date: April 25–26, 2003
Meeting moderator: Dr. John Mugabe, New Partnership for Africa's Development Science and Technology Forum

Day 1

0830–0900   Welcome and introductions
            Presentation: Dr. Tobias Takavarasha, Food, Agriculture, and Natural Resources Policy Analysis Network
            Introductions: Participants

0900–1000   Objectives, Expectations, and Ground Rules
            Presentation: Prof. Joachim von Braun, International Food Policy Research Institute
            Open plenary discussion: Moderated by Dr. Mugabe

1000–1030   Tea/coffee break

1030–1230   Agricultural Biotechnology and GMOs in Southern Africa: A Regional Synthesis
            Presentation: Dr. Doreen Mnyulwa and Julius Mugwagwa, Biotechnology Trust of Zimbabwe
            Open plenary discussion: Moderated by Dr. Mugabe

1230–1400   Lunch break

1400–1600   Dealing with Complex Public Disputes: Multiple-Stakeholder Approaches, Negotiation, and the Practice of Consensus Building
Presentation: Ms. Michele Ferenz, Consensus Building Institute
Open plenary discussion: Moderated by Dr. Mugabe

1600–1630   Tea/coffee break

1630–1800   The Road Ahead: Where We Might Go from Here
Open plenary discussion: Moderated by Dr. Mugabe

Day 2

0830–0930   Overview of day 1 and preparation for day 2 activities
Presentation: Dr. Mugabe
Open plenary discussion: Moderated by Dr. Mugabe

0930–1800   Day 2 activities
Plenary and group-based discussions of selected topics
Selection of steering committee members

1830–2000   Meeting of Steering Committee

Steering Committee Meeting Notes

Present: Fred Kalibwani, John Mugabe, Julius Mugwagwa, Norah Olembo, Cathy Rutivi (representing Tobias Takavarasha), Unesu Ushewokunze-Obatolu, and Klaus von Grebmer. *Secretary:* Jenna Kryszczun

Committee Membership

The committee was selected before the closing of the policy dialogue on biotechnology, agriculture, and food security in southern Africa on April 26, 2003. The committee members are

- John Mugabe (*chair; ex-officio*), Executive Secretary, Science and Technology Forum, New Partnership for Africa's Development

- Fred Kalibwani, Advocacy Officer, PELUM Association

- Julius Mugwagwa, Research Coordinator, Biotechnology Trust of Zimbabwe

- Norah Olembo, Managing Director, Kenya Industrial Property Institute

- Unesu Ushewokunze-Obatolu (*interim*), Deputy Director General (Research), AREX, MOLARR and Vice-Chair, Southern African Development Community (SADC) Advisory Committee on Biotechnology [interim member until SADC Advisory Committee elects a representative]

- Tobias Takavarasha (*ex-officio*), Chief Executive Officer, Food, Agriculture, and Natural Resources Policy Analysis Network (FANRPAN)

- Klaus von Grebmer (*ex-officio; interim*), Director, Communications, International Food Policy Research Institute (IFPRI) (interim member until Steven Were Omamo, Research Fellow/Network Coordinator, IFPRI, joins)

- Open

- Open

Meeting Agenda
The committee developed and approved the following agenda for the meeting:

1. Selection of chairperson

2. Interim communications strategy

3. Roles

4. Tasks (next steps)

5. Nomination and appointment of other two members

Selection of Chairperson
The committee elected Dr. Mugabe as its chair.

Interim Communications Strategy
It was decided that IFPRI would handle the communications aspects of the committee and develop a communications strategy. This would involve (a) setting up a listserv for committee members (within one month) and (b) ensuring that the committee members are networked electronically, that is, if there is an issue that

needs to be addressed, IFPRI will ensure that all members are networked so they are able to respond to that particular issue.

An electronic platform for the dialogue was discussed. Dr. von Grebmer indicated that a Web site could be set up that would have a section that was open to all, which would contain papers, proceedings of dialogues, and other information and links to each of the members' Web sites, as well as a section that was closed and open only to committee members, which could serve as a means of communication, sharing documents, and editing documents among committee members. Committee members expressed interest in having an electronic platform where papers could be available and comments posted in order to have electronic dialogues. The idea of posting a bibliography of information on biotechnology in Africa was also suggested, as was the posting of the tables in Dr. Mugwawa's presentation for updating by visitors to the Web site.

First Dialogue Report and Accompanying Letter

The committee asked the conveners of the meeting (IFPRI-FANRPAN) to prepare the synthesis report of the proceedings of the policy dialogue, to be reviewed by the chair of the dialogue, Dr. Mugabe. The conveners were also asked to prepare an accompanying letter, which will be sent to key actors, communicating the decisions and recommendations of the dialogue including the formation and composition of the committee. The letter should be signed by IFPRI-FANRPAN. Initial suggestions for recipients of the report/letter include the SADC, the SADC Advisory Committee on Biotechnology chairperson, the African Union, and the African Biotechnology Stakeholders' Forum. It was decided that committee members will develop a list of additional institutions (with contact names and information) that they would like to target and people that they would like to subsequently engage in the dialogue process. The list is to be sent to the conveners by e-mail. The committee requested that the letters and reports be sent out within one month (the week of May 26) to the developed list as well as to all participants in the dialogue.

Roles

Secretariat. In terms of secretariat-type activities, the committee decided to leave it to IFPRI and FANRPAN to discuss.

Participation of committee members in meetings and processes. The committee decided that if, as a member of the committee, a person is requested to fit into a process, the member shall have the flexibility to do so; however, the role of the member must be agreed upon so that he or she is not seen as being advisory or representative of the dialogue as a whole. Committee members should be considered a resource, to take information to those processes if and when required. It was proposed that the chair consider participating in discussions of the SADC Advisory Committee, depending on the response that this committee receives from the SADC Committee. It was noted that this would be considered upon request, and that the report could be presented, along with some of the key issues. It was recommended that the letter accompanying the report to the SADC committee suggest that the dialogue is open as a resource for that process.

Tasks

Calendar of key events. The committee agreed that a shared calendar of events would be useful. Dr. von Grebmer will design and set up the calendar within the next two months, and committee members will fill the calendar with important dates and events. It was discussed that the following key events would be milestones for the committee to target for the dialogue:

1. United States Agency for International Development Conference on Biotechnology—December 2003 (in Chile)

2. Convention on Biodiversity—April 2004 (in Latin America)

3. IFPRI Conference on Food and Nutrition Security—April 2004 (in Uganda)

Two other events to target were mentioned—the African Union summit in July 2003 and the Council of Ministers in August 2003, but given that the dates are so near, the effort would be to communicate that the dialogue and committee have been established.

Next dialogue. The committee agreed that there would be at least one more dialogue by the beginning of December 2003 and that it should target the above three noted processes. When the original timeline from the concept note was raised, the committee decided to consult among themselves in the coming months, particularly involving Dr. Omamo, about whether in fact they should hold another dialogue before December and if there is capacity and resources to organize the dialogue within that time. Concern was raised over the time between dialogues, and a need to keep up the momentum of the process was expressed. The committee agreed to propose to have two dialogues, one in September and one in January, and they will explore with IFPRI, particularly regarding funding and capacity.

It was suggested that perhaps one subregional activity could be held before another large dialogue in December, for example, a possible side event at the ministerial meeting in July. The committee decided to consult among themselves on key events on which the dialogue should focus and to also get input from Dr. Omamo.

Regarding the ministerial meeting, it was raised that the SADC Advisory Committee on Biotechnology will be presenting concept notes and working papers at the meeting to bring forward information that would help the ministers with formulation of policies and legislation. It was mentioned that it would be very instructive if the dialogue could inform the policy suggestion process to people who were assigned to undertake this work. The committee decided to ensure that the final report of the first dialogue as well as the working papers are formally submitted to that event in addition to a four- or five-page brief that offers highlights of the issues raised during the dialogue, flagging key issues while also indicating that the dialogue is available as a resource (noting the Web page if possible).

On the basis of consultation, the committee will talk with the secretariat regarding the capacity to organize a roundtable around the ministerial meeting in August. The committee agreed to have an electronic exchange to agree on a possible theme for that meeting, one that will influence the ministerial discussion. If resources are available and there is a need for another set of papers to be commissioned, the committee will work in consultation with the secretariat to collectively determine the terms of reference, while the secretariat will be left to commission the papers. The terms of reference for the second dialogue will be developed by the secretariat and the committee electronically.

Procedures for invitations to dialogues. The committee decided to develop a body of procedures for multistakeholder dialogues that would discuss the invitation procedure for dialogues. It was acknowledged that there may be some sessions that are open, but others that the committee may decide not to be open.

Production of report and brief. In summary, IFPRI will facilitate the production of a synthesis report and a brief coming out of the dialogue to be ready by mid-June. The brief will be drafted by IFPRI-FANRPAN and sent to committee members for review. It should contain a summary of the dialogue and key issues of the working papers, and note the key players, information about what the dialogue is, and key issues that the committee wishes to flag for which the dialogue may be available as a key resource for gathering information.

Committee meeting notes. Ms. Kryszczun was asked to send the committee meeting minutes as well as contact information for all committee members to the group by the beginning of the week of May 5.

Committee mandate. The committee decided to discuss via e-mail the development and interpretation of its mandate. The mandate will be taken to the next dialogue so it may be renewed.

Next committee meeting. The committee suggested meeting a day or two before the next dialogue.

Nomination and Appointment of Other Two Members

The committee decided that the two open slots on the committee will be filled at the next dialogue. It was proposed that several individuals from the sectors of consumer groups and farmers be invited to the next dialogue to ensure that there is an open process to nominate and appoint the additional two members. Committee members will be proactive in assisting the secretariat in identifying those to invite.

Closing

The Chair closed, urging the committee members to be actively engaged in the coming weeks. He noted that expectations emerging from the dialogue are relatively high, and the committee needs to ensure that those expectations are met.

Contributors

Mike Adcock, Sheffield Institute of Biotechnological Law and Ethics, Law Department, University of Sheffield, United Kingdom

Michele Ferenz, Consensus Building Institute, Boston, Massachusetts, USA

Julian Kinderlerer, Sheffield Institute of Biotechnological Law and Ethics, Law Department, University of Sheffield, United Kingdom

Jenna Kryszczun, International Food Policy Research Institute

David Matz, University of Massachusetts, Boston, USA

Doreen Mnyulwa, Regional Agricultural and Environmental Initiative–Africa and Biotechnology Trust of Zimbabwe

Moono Moputola, Southern African Development Community Hub, Harare, Zimbabwe

Julius Mugwagwa, Biotechnology Trust of Zimbabwe, Harare, Zimbabwe

Norah Olembo, University of Nairobi, African Biotechnology Stakeholders Forum, and Biotechnology Trust Africa

Steven Were Omamo, International Food Policy Research Institute

David Pelletier, Cornell University, Ithaca, New York, USA

Unesu Ushewokunze-Obatolu, Zimbabwe Department of Veterinary Public Health

Klaus von Grebmer, International Food Policy Research Institute

Index

Page numbers for entries occurring in figures are suffixed by an *f*; those for entries in notes by an *n*, with the number of the note following; and those for entries in tables by a *t*.

AATF. *See* African Agricultural Technology Foundation
ABSF. *See* African Biotechnology Stakeholders' Forum
Accountability, 53, 55, 61
Action maps, 67
Adulteration of foods, 121–22, 123, 126, 143
Advisory Committee on Ethics and Biotechnology in Animals (Netherlands), 105
Advocacy science, 44
Affected peoples, 42, 62
AfricaBio, 231
African Agricultural Technology Foundation (AATF), 180
African Biotechnology Stakeholders' Forum (ABSF), 35, 183, 232, 250–51, 274
African Center for Technology Studies, 232
African Growth and Opportunity Act, 193
African Intellectual Property Organization, 176
African Model Law, 178–79, 217, 236, 254, 256
African Regional Intellectual Property Organization (ARIPO), 176, 182
African Union (AU), 3, 255, 274

African Union Summit, 3
Agency for International Development, U.S., 181
Agenda 21, 79
Agreement on Technical Barriers to Trade (TBT), 189
Agreement on the Application of Sanitary and Phytosanitary (SPS) Measures, 81, 159, 189
Agreement on the Trade Related Aspects of Intellectual Property Rights (TRIPS), 90–91, 93, 94, 177, 179, 183, 189, 217; aim of, 90; Regional Policy Dialogue on, 236
Agricultural production: of GMO crops, SADC, 193–94; of GMO crops, world, 188–89; in SADC, 192
AIDS. *See* HIV/AIDS
Algeria, 74
Allergens, 88, 105, 143; FDA policy on, 123, 124, 127–28, 129; health status and, 139, 144; uncertainty factor and, 148t, 149; in U.S. versus African diet, 137–38
Alliance for Bio-Integrity v. Shalala, 130
Amino acid sequences, 128
Andersen, I.-E., 55

282   INDEX

Andhra Pradesh, India, 52–55
Angola, 16t, 20t, 27t, 174t, 231
Animal diseases and pests, 161
Animal feed, 137, 138, 190, 191, 195
Animal welfare, 104–5
Antibiotic resistance marker genes, 85–86
Antibiotics, 14
Area/neighborhood forums, 64
Argentina, 21, 82, 173, 188, 191, 196
ARIPO. *See* African Regional Intellectual Property Organization
Artificial insemination, 230
Ascorbic acid, 73
Ashby Committee, 72–73
Asia, 102, 104, 173
Asilomar, California, 72, 75
Association for Strengthening Agricultural Research in Eastern and Central Africa, 183
AU. *See* African Union
Australia, 21, 83, 173, 191
Autonomy, 100–101
Aventis, 191
Awareness. *See* Policymaker awareness; Public awareness

Bacillus thuringiensis (*Bt*), 261; cotton, 22–23, 194, 195, 231, 250; maize, 2, 138, 140, 141f, 143, 181, 194, 231, 251; toxin of, 128, 144
Backcrossing, 123, 124, 129
Bananas, 32, 196
Beans, 104; Enola, 93; kidney, 136; yellow, 92–93
Beef, 160, 192, 195–96
Beef and Veal Protocol, 192, 196
Benefits: just distribution of, 101–2; potential, 139; risks versus, 97–99
Berne Convention, 178t
Best alternative to a negotiated agreement, 240

Bilateral agencies, 42
Biological diversity, 73, 77–78, 79, 87, 190
Biological nitrogen fixation, 230
Biophysical sciences, 3–5, 202
Bioprospecting, 92–93
Biosafety Committee (Zambia), 231
Biosafety policy, 11, 32–33, 71, 72–82, 157–71, 183, 200, 201, 208, 215; background to, 24–26; challenges to, 166–67; consultation on, 34–35; draft of proposed framework for, 162–65t; financial resources and, 169–70, 219–20; framework for, 160–66, 211; historical perspective on, 72–75; international status of, 159; promoting, 209–13; public involvement in, 167–69; recommendations for, 170–71; Regional Policy Dialogue on, 234–36, 249–53, 254–55, 257, 259t, 263–64; status of, 16–17, 20–22, 159–60, 174t; trade and, 159, 161, 167, 213–14, 235, 253, 264. *See also* Environmental issues; Food safety policy
Biotechnology: defined, 12; generations of, 13–14; gradient of, 14, 15f; status of development and use, 16–19t; status of policy, 174t; traditional, 12
Biotechnology Alliance (Namibia), 249
Biotechnology Association (Zimbabwe), 231
Biotechnology Science Coordinating Committee (BSCC), 145
Biotechnology Trust of Africa, 183, 232
Biotechnology Trust of Zimbabwe (BTZ), 15, 30, 33, 36, 183, 230, 231
Biowatch (South Africa), 231
Botswana: biosafety policy in, 20t, 160, 174t; biotechnology development and use in, 16t; biotechnology regulations in, 77, 231; Cartagena Protocol ratified by, 74; public awareness in, 27t; trade in, 192, 193, 195
Bottom-up participatory process, 35

INDEX   283

Boundaries and context, values regarding, 142, 144, 146
Brazil, 35
British Medical Association, 84, 108n16
Broad-spectrum profiling, 150, 211
Brundtland Report, 99–100
BSCC. *See* Biotechnology Science Coordinating Committee
Bt. See Bacillus thuringiensis
BTZ. *See* Biotechnology Trust of Zimbabwe
Burkina Faso, 74
Bush, G. W., 191
Business stakeholders, 42

Calorie intake, 103–4, 116, 137, 152n2, 211
Cambridge University, 14
Cameroon, 74
Camp David talks, 38
Canada, 21, 34, 41, 80, 81, 105, 173, 181; biotechnology regulations in, 77; production of GMO crops in, 188; public opinion in, 82, 83; science-based decisions in, 87; trade with, 191
Canola, 88, 188
Capacity building, 215–16, 244, 254, 262; biosafety policy and, 169, 213; intellectual property rights and, 183–84; recommendations on, 30–32
Capacity constraints, 47–48, 205
Caribbean, 102
Carlson, C., 42
Cartagena Protocol on Biosafety, 11, 17, 77, 79, 82, 84, 87, 159, 210; countries ratifying, 74; historical background on, 73–74; objective of, 107n12; on openness, 102–3; on public involvement, 24–25, 86; Regional Policy Dialogue on, 235, 252, 254; trade policy and, 161, 190, 214
Cash crops, 193, 195

Cassava, 32, 104, 136, 181, 192
CBD. *See* Convention on Biological Diversity
Central Africa, 104
Centre de Coopération Internationale en Recherche Agronomique pour le Développement (CIRAD), 181
Cereals, 125, 136, 181, 192, 193, 195
CGIAR. *See* Consultative Group on International Agricultural Research
Chickpeas, 125, 136
Children, 42, 147, 150, 218, 219
Chile, 191, 196
China, 21, 188, 237
Christian Aid, 84, 85, 108n16
Chymosin, 73, 107n6
CIMMYT. *See* International Maize and Wheat Improvement Center
CIRAD. *See* Centre de Coopération Internationale en Recherche Agronomique pour le Développement
Citizens' juries, 35, 52–55, 66
Citizens' panels, 64
Civil society, 1, 7, 40, 47, 201, 204, 205, 208, 227; biosafety policy and, 166; information sharing in, 207; stakeholder status of, 42; World Commission on Dams and, 59
Cloning, 72, 73, 230
Codex Alimentarius, 159
Coffee, 193, 195, 196
Cohen, J. I., 160
Colombia, 32, 191
Commercial confidentiality, 26–28, 216, 232, 235, 236
Commercialization of biotechnology, 15, 22–23, 76–77, 160, 175, 200, 230
Commission of the European Communities, 80
Community issues groups, 64–65
Compositional substantial equivalence, 126
Comstock, G., 102

284 INDEX

Conflict assessment, 42–44
Conflict resolution, 240–41
Consensus building, 37–69; determining point of, 47; joint fact-finding process in, 46f, 207; phases of, 49f; recommendations on, 202, 203, 204, 205–6, 207; Regional Policy Dialogue on, 9–10, 227, 240, 244, 246. *See also* Multistakeholder processes
Consensus Building Institute, 240
Consensus conferences, 65
Consensus participation, 65
Consultation, tools for, 34–35
Consultative Group on International Agricultural Research (CGIAR), 181, 182
Consumer choice policy, 11, 103, 113–52; options and trade-offs in, 140–51; Regional Policy Dialogue on, 227, 229, 247, 262. *See also* Food safety policy
Consumers International, 232
Contingent agreements, 44
Controversy, 10, 24; on intellectual property rights, 179–80; on trade policy, 190–92
Convention on Biological Diversity (CBD), 12, 17, 73, 74, 77, 78, 159, 225; intellectual property rights and, 93, 94–95; trade policy and, 190
Coordinated Framework for the Regulation of Biotechnology, U.S., 74–75, 76t
Corn, 188, 191, 192. *See also* Maize
Corn Belt, 14
Cornmeal, 188
Costa Rica, 102
Cotonou Agreement, 192, 196
Cotton, 188, 191, 261; *Bt*, 22–23, 194, 195, 231, 250; trade in, 193, 194, 195
Cottonseed cake, 188t, 195
Cottonseed oil, 188t, 195
Council of Europe Parliamentary Assembly, 73, 86, 89–90, 95–96

Cowpeas, 104
Crisis of implementation, 242
Cruciferae, 136
Crustacea, 127
CRY9C, 138
Cry genes, 181, 182
Cucumbers, 125
Cucurbiticin, 125, 136
Cultural factors, 28, 135–37, 233
Cyanoglycosides, 125, 136, 149

Dalit caste, 54
Dams. *See* World Commission on Dams
Danish Board of Technology (DBT), 55–56, 58
Danish Center of Urban Ecology, 58
Dates, 193
Debates, 66, 257
Decision EX.CL, 3
Decision making: participatory, 61, 62; science-based, 86–89
Decisionmaking bodies, 44–48
Deliberative opinion polls, 66
Democratic Republic of Congo, 16t, 27t, 231
Denmark, 37, 55–59
Deoxyribonucleic acid. *See* DNA
Department for International Development (DFID)–India, 55
Department of Agriculture, U.S. (USDA), 77, 131
Dickson, D., 29
Dignity, 100–101
Directive 2001/18/EC, 190
Djibouti, 74
DNA, 14, 78, 230. *See also* Recombinant DNA technology
DNA bar code, 190
Doha conference, 93–94
Dolly (cloned sheep), 230

INDEX 285

Dow Agro Sciences, 181
Droughts, 29, 97–98, 113, 161, 192, 231, 239, 253
Drought-tolerant crops, 32, 114
Dummy field experimentation, 250
Dumping, 103
Dupont, 181

Earth Summit, 42, 241, 242
Eastern Africa, 30–31, 33, 160, 174t, 175
Educational tools, 35–36
Efficiency, 48, 61, 204
E-Forum on Participatory Processes for Policy Change, 52–55
Eggs, 127
Egypt, 38, 74, 191
Electronic democracy, 66
El Salvador, 191
Embryo transfer, 230
Encyclopaedia of the Atmospheric Environment, 100
Engagement issues, 53, 54–55
Enola beans, 93
Environmental issues, 34, 71, 73, 166, 168, 190, 199, 218; interface with other issues, 29; liability limitations and, 29; multistakeholder processes in, 41, 44, 48; policy on, 77–78; public opinion on, 85; Regional Policy Dialogue on, 229, 264; regulations on, 76; rights-based approach to, 37, 59–63; stakeholders in, 167; trade and, 99, 214. *See also* Biosafety policy; Precautionary principle; Sustainable development
Environmental Protection Agency (EPA), U.S., 41, 138
Environment and Development Activities (Zimbabwe), 29
Enzymes, 14

EPA. *See* Environmental Protection Agency, U.S.
Epitopes, 128
Equity, 61
Ethics, 10, 11, 202–4, 208; intellectual property rights and, 89–95, 184; specific issues in, 95–105
Ethiopia, 74, 174t, 175t
EU. *See* European Union
Eurobarometer surveys, 82, 83
Europe, 29, 41, 231; biosafety policy and, 73–74, 230; biotechnology regulations in, 77; intellectual property rights and, 91–92; optimism about technologies in, 84f; public opinion in, 82, 83, 84–86; trade with, 167, 194–95, 214
European Novel Food Regulation, 78
European Patent Convention, 92
European Patent Office, 92
European Union (EU), 3, 29, 71, 72, 73, 78, 105, 192, 195–96; biosafety policy and, 160, 210, 253; biotechnology policy in, 233; controversies over trade policy of, 190–92; science-based decisions in, 87
Evidence issues, 53–54
Exploitation, 103–4
Export credit guarantee agencies, 42
Exports, 85, 93, 113, 150–51, 160, 189, 191, 192, 212, 253; facilitating, 213–15; growth in, 193; liability limitations and, 29; policy issues and trade-offs, 194–96

Facilitators and facilitation, 39–40, 48, 50
Fairness, 48, 204, 242
Famines, 29
FanMeat, 195
FANRPAN. *See* Food, Agriculture, and Natural Resources Policy Analysis Network
FAO. *See* Food and Agriculture Organization

Farmers. *See* Smallholder farmers
FDA. *See* Food and Drug Administration, U.S.
Federal Register, 115
Ferenz, M., 240–44, 245, 246, 272
Fermentation technology, 14, 32, 230
Fertilizers, 239
Field tests, 79, 82, 87, 250
Figs, 193
Financial Gazette, 76–77
Financial resources: biosafety policy and, 169–70, 219–20; creating sustainable, 219–21; Regional Policy Dialogue on, 237–38, 255
First generation of biotechnology, 14
Fish, 127
Floods, 161, 192, 253
Flowers, 193
Focus groups, 64, 66
FOIA. *See* Freedom of Information Act
Folate, 114
Folic acid, 132
Food additives, 121, 122, 126
Food, Agriculture, and Natural Resources Policy Analysis Network (FANRPAN), 1, 219–20. *See also* Regional Policy Dialogue
Food Agriculture and Natural Resources Sector, 160–61
Food aid, 2, 35, 84, 113, 166, 187, 191, 209, 213; interface with other issues, 29–30; Regional Policy Dialogue on, 253
Food and Agriculture Organization (FAO), 89, 98, 104, 125, 159, 182, 259
Food and Drug Administration (FDA), U.S., 11, 114, 115, 118–37, 143, 144, 145, 150, 151, 210, 251; action levels of, 122; effectiveness of regulations, 130t; failure to address cultural issues, 135–37; legal framework for, 118–23; proposed rules for 2001, 131–33; scientific issues and, 123–24; timeline for key policy events, 118, 119–20t
Food crises, 1–2, 231
Food, Drug and Cosmetic Act, 121–22, 131
Food insecurity. *See* Food security/insecurity
Food processing, 135–37
Food safety policy, 11, 113–52, 218; options and trade-offs in, 140–51; Regional Policy Dialogue on, 262; southern African context, 135–39; U.S. versus southern Africa context, 114, 116–18, 137–39. *See also* Biosafety policy; Food and Drug Administration, U.S.
Food security/insecurity, 98, 161, 199, 200, 214, 220, 252, 260t, 261; developing strategy for, 216–19; food safety policy and, 147, 149–50; intellectual property rights and, 179; uncertainty factor and, 147, 148t
France, 21, 84
Freedom of Information Act (FOIA), 127
Free trade area (FTA), 187, 192, 193
Friends of the Earth, 84, 108n16
Frontiers, 168
Fruit juices, 193
FTA. *See* Free trade area
Functional substantial equivalence, 126
Fungus-resistant crops, 114
Future search conferences, 66–67

Gambia, 74
GAO. *See* Government Accounting Office
GEF. *See* Global Environment Facility
Generally regarded as safe standard. *See* GRAS standard
General welfare, principle of, 97–99
Generations of biotechnology, 13–14
Genes, 14, 86; cry, 181, 182; terminator, 182; trans, 124, 127
"Genetically Modified Crops: The Ethical and Social Issues" (report), 95

Genetically modified organisms (GMOs): defined, 107n11; four-year ban on, 190–92; global and regional production trends, 20–21; overview of use in SADC, 23; SADC production of, 193–94; world production and trade in, 188–89
Genetic Resources International (GRAIN), 180
Geneva, Kenya, 177
Germany, 251
Ghana, 74
Global Environment Facility (GEF), 26, 30, 77, 159, 161, 169
Glucosinolates, 88, 136
Glycol-alkaloid accumulation, 88
GMOs. *See* Genetically modified organisms
Government, 42, 48, 211–12, 234, 244
Government Accounting Office (GAO), 126
Gradient of biotechnology, 14, 15f
GRAIN. *See* Genetic Resources International
GRAS (generally regarded as safe) standard, 121, 122, 123, 126, 131
Green Foundation, 58
Greenpeace, 84, 108n16, 244
Green Revolution, 102, 116, 173
Groundnuts, 104

Hague Agreement, 178t
Halal foods, 98
Hallman, W. K., 72, 83
Harare, Zimbabwe, 176, 182
Harmonization, 2, 33, 209–10, 213, 216; Regional Policy Dialogue on, 227, 228, 234, 256, 257, 262; trade and, 214
Harvest, A, 183
Health status, 29, 117, 139, 144, 209, 218; biosafety policy and, 167; ethical consideration of, 98; intellectual property rights and, 94; trade policy and, 194. *See also* Morbidity
Hemolytic agents, 136
Herbal medicines, 92
Herbicides, 73, 102, 239
Herbicide-tolerant crops, 86, 193
HIV/AIDS, 94, 117, 139, 169, 211, 218, 251, 262
Holmes, T., 64
Honduras, 191
Humanities, 3–5
Hunger, 173, 218
Hybrid crops, 14

ICRISAT. *See* International Maize and Wheat Improvement Center
IIED. *See* International Institute for Environment and Development
Imports, 23, 157, 194, 195, 212, 213; bans on, 191, 192; Cartagena Protocol on, 25; growth in, 193; Regional Policy Dialogue on, 253
Independent advisory committees, 35
Independent public inquiries, 35
India, 32, 35, 173, 237; electronic multistakeholder dialogue in, 37, 52–55; Green Revolution in, 102; neem tree in, 92, 108n20
Indigenous people, 42
Indigenous plants, 236–37
Information sharing, 38, 206–8; Regional Policy Dialogue on, 249–56; tools for, 35–36
Innovative development, 67
Insecticides, 73
Insect-resistant crops, 114, 181, 193, 250
Insect Resistant Maize for Africa (IRMA), 181, 182
Insertional mutagenesis, 123, 124, 127, 131, 135
Institute of Development Studies, 53
Insurance, 143

Integrative bargaining, 38
Integrity, 100–101
Intellectual property rights, 11–12, 71, 89–95, 173–85; abuse of, 103; benefits of, 216–17; conceptual framework for, 180–83; controversy over, 179–80; disadvantages of, 91; extension of, 90; importance of, 175–77; issues and policy needs, 174–75; participation in agreements on, 178t; policy trade-offs for, 180–83; recommendations on, 184–85; Regional Policy Dialogue on, 229, 233, 239, 259t, 261, 262; status of knowledge on, 177–79; status of laws on, 175t; trade and, 194
Interactive panels, 64
Intermediate Technology Development Group, 35
International associations, 42
International Center for Maize and Wheat, 194
International Center for Tropical Agriculture, 93
International Convention for the Protection of New Varieties of Plants, 91
International Food Policy Research Institute (IFPRI). *See* Regional Policy Dialogue
International Institute for Environment and Development (IIED), 41, 52, 53
International Maize and Wheat Improvement Center (CIMMYT; ICRISAT), 181, 263
International Monetary Fund, 40
International Seed Federation, 188
International Service for National Agricultural Research, 15, 255
International Service for the Acquisition of Agro-biotech Applications, 21, 181
International Treaty for the Protection of Plant Genetic Resources, 93
International Treaty on Plant Genetic Resources for Food and Agriculture (ITPGR), 182

International Union for the Protection of New Varieties of Plants (UPOV), 91, 177–78, 217
IRMA. *See* Insect Resistant Maize for Africa
Iron (dietary), 114, 150, 219
Israel, 38
Issue forums, 67
ITPGR. *See* International Treaty on Plant Genetic Resources for Food and Agriculture

Jaeger, B., 55
Jakarta, Indonesia, 73
Johannesburg, South Africa, 9, 10, 180, 201
Joint fact-finding (JFF) process, 44, 45f, 46f, 207
Just distribution, 101–2
Justice, principle of, 96–97, 203

Kalibwani, F., 272
Kandawa-Schulz, M., 249, 252–53
Kenya, 32, 181, 263; biosafety policy in, 174t, 250–53; biotechnology regulations in, 77; Cartagena Protocol ratified by, 74; intellectual property rights in, 175t, 177, 178, 182
Kenya Industrial Property Institute, 182
Kidney beans, 136
Kosher foods, 98
Kryszczun, J., 272, 276
Kuiper, H. A., 133
Kwazulu-Natal Province, 22, 193

Labeling, 23, 85, 151, 211; FDA policy on, 118, 143, 144; Regional Policy Dialogue on, 252–53; trade policy and, 190–91, 195–96
Laboratory tests, 15, 82, 181, 250, 251
Language barriers, 28, 168–69
Lathyrogens, 125, 136
Latin America, 102, 104

INDEX 289

Lectins, 125, 136, 149
Legal framework: enabling, 34; FDA, 118–23; for trade policy, 189–90
Legumes, 125, 127, 136, 149
Lentils, 104, 181
Lesotho: biosafety policy in, 20t, 174t; biotechnology development and use in, 17t; biotechnology regulations in, 77, 231; Cartagena Protocol ratified by, 74; intellectual property rights in, 175t; public awareness in, 27t; trade in, 193, 195
Liberia, 74
Licensing systems, compulsory, 92, 94
Life sciences, 158, 231
Literacy rates, 28, 167, 169
Livestock, 193, 195
Local Agenda 21, 69
Local authorities, 42
Local knowledge, 40
Luhanga, B., 253, 255–56

Madagascar, 74
Madrid Agreement, 178t
Maize, 29, 30, 84, 102, 181, 187, 211, 250, 252, 253; acreage devoted to, 173; *Bt,* 2, 138, 140, 141f, 143, 181, 194, 231, 251; calories provided by, 103, 152n2; commercialization of, 23; production of, 188t, 192; Starlink, 138; trade in, 188t, 193, 194, 195. *See also* Corn
Makhathini Flats, 22
Malaria, 94
Malawi, 2, 30, 35, 200, 232, 253; biosafety policy in, 16, 21t, 160, 169, 174t, 209; biotechnology development and use in, 15, 17t; biotechnology regulations in, 231; intellectual property rights in, 175t; public awareness in, 27t; trade in, 192
Mali, 74

Malnutrition, 97, 211, 218, 219, 251; factors responsible for, 173; food safety policy and, 117, 139, 147, 148t, 149–50
Mandaza, I., 167
Maputo, Mozambique, 2
Marker-assisted selection, 230
Market mechanisms, 143
Marrakesh Agreement, 189
Mauritius, 200, 232; biosafety policy in, 21t, 174t; biotechnology development and use in, 15, 17t; biotechnology regulations in, 231; Cartagena Protocol ratified by, 74; intellectual property rights in, 175t; public awareness in, 27t; trade in, 192
McClean, M. A., 161
MCM. *See* Multicriteria mapping
Media, 36, 85, 245
Mediators and mediation, 40, 48, 50
Medline, 130
Metanarratives, 5
Mexico, 21, 83, 92–93, 181, 191
Middle East, 102
Milk, 127
Millet, 104, 192
Mining, Minerals, and Sustainable Development Project, 40, 41
Mnyulwa, D., 160, 271
Moderators, 50, 66, 245
Modernism, 5–6
Molecular biology, 32
Molecular diagnostics and markers, 230
Mollusks, 127
Monopoly status, 101, 179
Monsanto, 181, 191
Montreal, Canada, 73
Morbidity, 147, 251. *See also* Health status
Movable options cards, 68
Mozambique, 191; biosafety policy in, 21t, 161, 174t; biotechnology applications in,

Mozambique (*continued*)
17t; biotechnology regulations in, 77, 231; Cartagena Protocol ratified by, 74; intellectual property rights in, 175t; public awareness in, 27t
MSDs. *See* Multistakeholder dialogues
Mugabe, J., 9, 223, 224, 225–26, 247, 249, 257–59, 265, 267, 269, 271, 272, 273, 274
Mugwagwa, J., 230–32, 238, 271, 272, 274
Multicriteria mapping (MCM), 67
Multidimensionality, 48–52
Multilateral agencies, 42
Multistakeholder dialogues (MSDs), 40–41, 246; alternatives to, 245; decisionmaking bodies in, 44–47; electronic, 37, 52–55
Multistakeholder processes, 1, 3, 10–11, 12, 37–69, 220; alternatives for design of, 64–69; challenges addressed by, 241–44; concept of, 37–48; contingency and monitoring plans, 39–40; criteria for dividing joint gains, 39; decisionmaking bodies and, 44–48; examples of, 52–63; expanding and sustaining, 201–6; fairness and efficiency in, 48; information sharing in, 38; invention of options for, 38–39; lessons learned from, 41–48; packaging of options for, 39; participatory planning in, 40–41; policy-focused approaches in, 41; Regional Policy Dialogue on, 9, 241–44; relevant parties in, 42–44; rights-based approach in, 37; scientific information in, 44. *See also* Stakeholders
Museveni, Y., 267

Nairobi, Kenya, 74, 180
Namibia, 25, 232; biosafety policy in, 21t, 160, 169, 174t, 249; biotechnology development and use in, 18t; biotechnology regulations in, 77, 231; Cartagena Protocol ratified by, 74; information sharing in, 249; intellectual property rights in, 175t; public awareness in, 27t; trade in, 192, 193, 195
National Academy of Sciences (NAS), U.S., 115, 134–35, 145, 146
National biosafety frameworks (NBFs), 24, 26
National Biotechnology Alliance (Zambia), 231
National Biotechnology Development Agency of Nigeria, 183
National Corn Growers Association, 191
National Council of Science and Technology Act, 175
National Institute for Scientific and Industrial Research (Zambia), 231
National Institutes of Health (NIH), U.S., 74, 75, 145
National Research Council (NRC), U.S., 115, 126, 127, 130, 132, 134–35, 136–37
Natural products, 96
NBFs. *See* National biosafety frameworks
Neem products, 92, 108n20
Negotiated rulemaking, 41
Neighborhood Initiatives Foundation, 68
NEPAD. *See* New Partnership for Africa's Development
Nestle UK, 108n17
Netherlands, 34, 105, 250
Neurotoxins, 136
New Partnership for Africa's Development (NEPAD), 3, 9, 224, 225, 232, 247, 256, 259, 262, 264, 265, 266, 267, 269
New Zealand, 191
NGOs. *See* Nongovernmental organizations
Nice Agreement, 178t
Niger, 74
Nigeria, 74, 102, 183
NIH. *See* National Institutes of Health, U.S.
Nongovernmental organizations (NGOs), 41, 53, 138, 143, 201, 205; influence on policy,

216; Regional Policy Dialogue on, 227, 244; as stakeholders, 42
Normative theories, 140, 142, 145–46, 203–4
North Africa, 102
North versus South, political myths in, 6–7
Novartis Foundation, 194
Novel foods, 77, 92, 93, 105, 108n14, 136
NRC. *See* National Research Council, U.S.
Nuffield Council on Bioethics, 95, 96–97, 202–3
Nutritional status, 139, 144, 147. *See also* Hunger; Malnutrition; Starvation

OECD. *See* Organization for Economic Cooperation and Development
Olembo, N., 250–53, 272, 273
Omamo, S. W., 269, 275, 276
Openness, 102–3
Orange Free State, 193
Ordre public, 72, 90
Organization for Economic Cooperation and Development (OECD), 41, 79, 88, 105, 125, 179
Organization of African Unity, 178–79, 217
Organization of Organic Producers and Processors (Zambia), 196
Our Common Future, 99–100
Outcomes: list of potential, 148t; values regarding, 142–43, 146
Oxfam, 195

Paarlberg, R., 160
Paris Union, 178t
Participatory decisionmaking, 61, 62
Participatory planning, 40–41
Participatory research and action (PRA), 67–68
Participatory rural appraisal, 67–68
Patent Cooperation Treaty, 178t

Patents, 89–93, 94, 108n18, 181, 182, 189; controversy over, 179; database of, 176–77; exclusion from, 90–91, 92, 106n1, 177; for genes, 86; Regional Policy Dialogue on, 236, 237
Peanuts, 127
Pelletier, D., 261, 268
Penicillin, 14, 230
People's Panel, 64
Peru, 191
Pesticides, 102, 108n20, 138, 187, 194, 239
Pest-resistant crops, 130, 158
Pharmaceuticals/drugs, 14, 73, 82, 92, 94
Phytate, 114, 132
Phytosanitary regulations, 196
Pigeon peas, 104
Pineapples, 193
Planning for real, 68
Plantain, 104
Playing God, 96, 105, 202–3
Pleiotropy, 123, 124, 127, 131, 135
POD-NERS, LLC, 92–93
Policy, 72–82; awareness of, 231–32; checklist of questions for, 48–52; design and implementation of, 215–16; disputes over, 1; environmental, 77–78; historical background on, 72–75; multistakeholder processes in, 41; poverty alleviation and, 218–19; Regional Policy Dialogue on, 225–26, 229, 232–33, 259–64; in response to public reactions, 24–30. *See also* Biosafety policy; Food safety policy; Intellectual property rights; Regulations; Trade
Policymaker awareness, 231–32
Political myths, 6–7
Politics: food safety policy and, 140–42; interface with other issues, 29–30
Poor: bias against, 104; increasing awareness in, 207

Population growth, 173
Positive theories, 140, 142, 204
Postmodernism, 5–6
Potatoes, 88, 104, 130
Potentialities, 67
Poverty alleviation strategy, 216–19
PRA. *See* Participatory research and action
"Prajateerpu: A Citizens Jury/Scenario Workshop on Food and Farming Futures for Andhra Pradesh," 52–55
Precautionary principle, 78–82, 146, 190, 191, 220–21
Private sector, 7, 42, 201, 234
Procedural fairness, 48
Proctor, J., 92–93
Prolamine, 132
Protease inhibitors, 125, 136
Protein, 127–28
Protoplast fusion, 123, 127
Pro-vitamin A, 85, 114, 132, 150, 219
Public awareness, 232, 239, 254; background, 24–26; building, 206–8; defined, 25; levels of, 27t; status of in SADC, 26; SWOT analysis of, 30
Public dialogue. *See* Public participation
Public disputes, 240–49
Public opinion, 82–86
Public participation: background, 24–26; in biosafety policy, 167–69, 211; challenges of, 26–30; defined, 25; tools for, 34–35

RAEIN-Africa. *See* Regional Agricultural and Environmental Initiative
Rapeseed, 188
Rapid rural appraisal (RRA), 68
Reasonable certainty of no harm standard, 122
Recombinant DNA (rDNA) technology, 72; FDA policy on, 123–24, 127, 129, 131–32; possible unpredictability of, 151–52

Recommendation 1213, 73
Red lines, 44
Regional Agricultural and Environmental Initiative (RAEIN-Africa), 169
Regional Agricultural and Environmental Network (Zimbabwe), 231
Regional Early Warning Unit (REWU), 192
Regional Policy Dialogue, 1, 8–10, 200, 201, 219–20, 223–69; links with other initiatives, 265; mandate of, 266–67; meeting notes, 272–77; nomination criteria for, 267; objectives, expectations, and ground rules, 224–29; program, 271–72; regional scope of, 265; regional synthesis, 230–39; reporting/coordination in, 266; role of, 265–66; timeline/benchmarks for, 267–68
Regional synthesis, 10, 13–36; on biosafety systems, 20–22; on public and policy responses, 24–30; recommendations based on, 30–33; in Regional Policy Dialogue, 230–39
Regulation 1829/2003, 78
Regulation 1830/2003, 78
Regulations, 30–31, 105–6, 209–13; basis for in life sciences, 158; interface with other issues, 30; Regional Policy Dialogue on, 231, 234–36; science and values in, 144–46; status of, 174t; strengthening, 150–51; systems for, 75–77; in the U.S., 74–75, 77. *See also* Policy
Religious issues. *See* Playing God
Representation, 53, 54, 244
Research, 34; capacity strengthening in, 215–16; enabling environment for, 32–33; intellectual property rights and, 183–84; investing in, 208–9; participatory, 67–68; recommendations on, 30–33; Regional Policy Dialogue on, 232–33, 238, 239, 261; status of in SADC, 14–17

INDEX 293

Research institutes, 42
Research panels, 64
REWU. *See* Regional Early Warning Unit
Rice, 83, 85, 102, 103, 132
Rights-based approach, 37, 59–63
Risk assessment, 11, 28, 33, 87–88, 159, 211, 251, 252; Cartagena Protocol on, 79; trade and, 196, 214
Risk management, 28, 33, 211; methods of, 143–44; trade and, 159, 214
Risks: benefits versus, 97–99; posed by dams, 60, 61, 62; potential, 158; Regional Policy Dialogue on, 239
River basin authorities, 42
Rockefeller Foundation, 15, 180, 261
Role groups, 57
Rome Declaration on World Food Security, 98
Royal Society of Canada, 87, 88
Royal Society of the United Kingdom, 88–89, 105
RRA. *See* Rapid rural appraisal
Rutivi, C., 272
Rwanda, 74, 77

SACU. *See* Southern African Customs Union
SADC. *See* Southern African Development Community
Safe Age, 231
Safety tests, 121, 129, 143, 211. *See also* Field tests; Laboratory tests
Sainsbury's, 196
Saline-resistant crops, 114
Sanitary measures, 196
Scenario workshops: in Denmark, 37, 55–59; in India, 52–55
SciDev.Net, 29
Science, 203, 230–32; advocacy, 44; biophysical, 3–5, 202; biosafety policy and, 169; decisions based on, 86–89; in FDA policy statement, 123–28; food safety and, 113–14, 140–46; interface with other issues, 29–30; life, 158, 231; modernism versus postmodernism on, 5–6; multistakeholder processes on, 44; normative approach and, 145–46; policymaker awareness of, 231–32; simplifying information, 28; social, 3–5, 202; stakeholders in, 42; variations in expertise, 230–31. *See also* Technology
Scoones, I., 52, 64
Second generation of biotechnology, 14
Seed Network, SADC, 263
Seeds, 188, 235, 236–37, 252, 261, 263, 264
Selective breeding, 12
Senegal, 74
Service user forums, 68
Seychelles: biosafety policy in, 22t, 174t; biotechnology development and use in, 18t; biotechnology regulations in, 231; Cartagena Protocol ratified by, 74; public awareness in, 27t
Silent pathways, 127
Sinai Desert, 38
Smallholder farmers, 227, 262; food safety policy and, 116; grievances of, 248; rights of, 235, 236–37; risks faced by, 158; as stakeholders, 42; trade and, 214
Social sciences, 3–5, 202
Social values. *See* Values
Soft pesticides, 108n20
Somaclonal variation, 123
Sorghum, 104
South Africa, 21, 35, 83, 173, 230–31, 232, 235, 239; biosafety policy in, 16, 22t, 160, 161, 174t, 209; biotechnology development and use in, 15, 18t; biotechnology regulations in, 76–77, 231, 234; Cartagena Protocol ratified by, 74; commercialization in, 15, 22–23, 76–77, 175, 200, 230;

South Africa (*continued*)
intellectual property rights in, 177; public awareness in, 27t; science in, 169; trade in, 192, 193, 194, 196
Southern African Customs Union (SACU), 193
Southern African Development Community (SADC): Advisory Committee on Biotechnology, 228, 253–56, 261, 265, 267, 274, 275, 276; agricultural production in, 192; biosafety framework for, 160–66; overview of GM use in, 23; Seed Network, 263; status of biosafety in, 20–22t, 159–60; status of biotechnology in, 16–19t; status of research in, 14–17; trade in, 192–93
Southern African Regional Biosafety Initiative, 169
Southern Africa Regional Biosafety Program, 232
South versus North, political myths in, 6–7
Soybean cake, 188t
Soybean oil, 188t
Soybeans, 104, 127, 132, 173, 188, 193, 196, 252
Spain, 21
SPS measures. *See* Agreement on the Application of Sanitary and Phytosanitary Measures
Squash, 125, 136
Stakeholder decision analysis, 68–69
Stakeholders: defined, 25; differing viewpoints, 3–5; in environmental issues, 167; identifying, 244; types of, 42; uninformed versus well-informed, 3. *See also* Multistakeholder processes
Staple foods, 137–39, 211
Starlink maize, 138
Starvation, 2, 173, 191

"Statement of Policy: Foods Derived from New Plant Varieties" (FDA), 118
Stem borers, 181
Stewardship, 105
Strengths, weaknesses, opportunities, and threats analysis. *See* SWOT analysis
Striga-resistant crops, 181
Substantial equivalence, 88–89, 125–26
Sugar, 192, 193, 195
Sugar beets, 104
Sugar cane, 104, 231
Sui generis protection, 91, 94, 177
Sustainable development, 61, 99–100, 161
Sustainable housing, 56–59
Swaziland: biosafety policy in, 22t, 174t; biotechnology development and use in, 19t; biotechnology regulations in, 231; intellectual property rights in, 175t; public awareness in, 27t; trade in, 192, 193
Sweet potatoes, 32, 104, 181, 250
SWOT (strengths, weaknesses, opportunities, and threats) analysis, 30, 31t
Syngenta, 181, 192

TA. *See* Technology assessment
Takavarasha, T., 228, 268–69, 272, 273
Tanzania, 2, 238; biosafety policy in, 174t, 234, 264; biotechnology development and use in, 19t; biotechnology regulations in, 231; *Bt* cotton cultivation in, 23t; Cartagena Protocol ratified by, 74; intellectual property rights in, 175t; public awareness in, 27t
Taste tests, 129
TBT agreement. *See* Agreement on Technical Barriers to Trade
Tea, 193, 195
Technology: development and transfer of, 259t, 260t; investment in, 232; modernism versus

INDEX   295

postmodernism on, 5–6; values regarding, 142. *See also* Science
Technology assessment (TA), 9
Terminator gene, 182
Tescos, 196
Theme groups, 57
Third generation of biotechnology, 14
Thompson, J., 52
Tissue culture techniques, 15, 32, 230
Tobacco, 193, 195, 264
Togo, 74
Tomatoes, 136
Top-down approaches, 40, 166–67
Toxicants, 88, 105, 143, 251; *Bt,* 128, 144; FDA policy on, 123, 124, 125–27, 129, 136; health status and, 144; uncertainty factor and, 148t, 149; in U.S. versus African diet, 137
Traceability, 190, 191, 195
Trade, 12, 187–97, 201, 209; biosafety policy and, 159, 161, 167, 213–14, 235, 253, 264; controversies over, 190–92; environmental concerns and, 99, 214; facilitating, 213–15; food safety and, 113; in GMO crops, world, 188–89; intellectual property rights and, 90–91, 176, 180; international legal framework for, 189–90; policy issues and trade-offs, 194–96; public participation and, 28–29; Regional Policy Dialogue on, 227, 229, 233, 257–58, 259t, 262, 264; in SADC, 192–93; website on, 197n1. *See also* Exports; Imports
Trade unions, 42
Traditional biotechnology, 12
Transgenes, 124, 127
Transparency, 24, 28, 33, 34, 42, 48, 67, 242
Tree nuts, 127
TRIPS. *See* Agreement on the Trade Related Aspects of Intellectual Property Rights

Trypanosomosis, tolerance to, 4–5
Tuberculosis, 94
Tunisia, 74

Uganda, 74, 77, 174t, 175t
UN. *See* United Nations
UNCED. *See* United Nations Conference on Environment and Development
Uncertainties, 39, 151, 208; comparison of, 147–49; Regional Policy Dialogue on, 239; values regarding, 142, 143–44, 146, 207
UNEP. *See* United Nations Environment Program
UNESCO. *See* United Nations Educational, Scientific, and Cultural Organization
Unilever UK, 108n17
Unintended effects of genetic breeding, 133t
United Kingdom, 34, 55, 64, 65, 95; biosafety policy in, 72–73, 74; biotechnology regulations in, 75; trade with, 196
United Nations (UN), 41, 42, 71, 74, 191, 234, 257
United Nations Conference on Environment and Development (UNCED), 40, 42
United Nations Educational, Scientific, and Cultural Organization (UNESCO), 100–101
United Nations Environment Program (UNEP), 25, 26, 30, 34, 77, 159, 161, 167, 169; International Guidelines of, 84; Regional Policy Dialogue on, 232, 249, 255
United Republic of Tanzania. *See* Tanzania
United States, 3, 11, 21, 23, 41, 71, 105, 113, 150–51, 203, 231, 251, 252; biosafety policy in, 210–11; biotechnology policy in, 233; biotechnology regulations in, 74–75, 77; food aid from, 2, 29, 191, 253; food safety policy in, 114, 116–18, 137–39; intellectual property rights in, 91, 92–93;

United States (*continued*)
leading GM crops in, 173; production of GM crops in, 188; public opinion in, 82, 83; science-based decisions in, 87; technology assessment in, 9; trade with, 190–91, 192, 196. *See also* Food and Drug Administration, U.S.
Universal Declaration on the Human Genome and Human Rights, 100–101
University of Sussex Institute for Development Studies, 263
Unnatural products, 96
UPOV. *See* International Union for the Protection of New Varieties of Plants
Urban ecology, 56–59
Uruguay, 191
USDA. *See* Department of Agriculture, U.S.
Ushewokunze-Obatolu, U., 253–54, 255, 272, 273

Vaccines, 73
Values, 140–46, 147, 207
Virus-resistant crops, 114, 181
Vision 2020, 52, 54
Visioning exercises, 69
Vitamin A. *See* Pro-vitamin A
Vitamin B12 deficiency, 132
Vitamins, 14, 114
Von Braun, J., 224–25, 228–29, 247, 256–57, 265, 268, 269, 271
Von Grebmer, K., 272, 273, 274, 275
Vulnerability, 100–101

Walker, S., 91, 95
WCD. *See* World Commission on Dams
Wheat, 102, 103, 127
White man's dilemma, 6
WHO. *See* World Health Organization
Wide-cross hybridization, 123
Windhoek, Namibia, 30
WIPO. *See* World Intellectual Property Organization
Women, 42, 54, 102, 147, 150, 204, 218, 219, 267
Workers, 42
World Bank, 41, 102, 250
World Business Council for Sustainable Development, 41
World Commission for the Environment and Development, 99–100
World Commission on Dams (WCD), 37, 41, 42, 59–63, 240
World Food Program, 29
World Health Organization (WHO), 88, 125, 159
World Intellectual Property Organization (WIPO), 177
World Summit on Sustainable Development, 180, 242
World Trade Organization (WTO), 28–29, 40, 81, 99, 189, 190, 215, 225, 246, 258, 259; biosafety policy and, 159, 213, 214; European Union and, 191; intellectual property rights and, 90–91, 93–94, 176
WTO. *See* World Trade Organization

Xanthophylls, 132

Yams, 32, 104
Yellow beans, 92–93

Zambia, 29, 30, 32, 191, 200, 228–29, 232, 253; biosafety policy in, 161; biotechnology development and use in, 15, 19t; biotechnology regulations in, 77, 231; *Bt* cotton cultivation in, 23t; Cartagena Protocol ratified by, 74; intellectual property rights in, 175t; public awareness in, 27t; public opinion in, 84; trade in, 192, 196

Zero-sum game, 39

Zimbabwe, 29, 30, 32, 35, 102, 181, 191, 200, 232, 245, 253, 263; biosafety policy in, 16, 160, 161, 174t, 209; biotechnology development and use in, 15, 19t; biotechnology regulations in, 76–77, 231; *Bt* cotton cultivation in, 23t; Cartagena Protocol ratified by, 74; intellectual property rights in, 175t; public awareness in, 27t; public opinion in, 84; science in, 169; trade in, 192, 193

Zinc (dietary), 114